DESIGNING PNEUMATIC CONTROL CIRCUITS
EFFICIENT TECHNIQUES FOR PRACTICAL APPLICATION

Bruce E. McCord
The Aro Corporation, Bryan, Ohio

MARCEL DEKKER, INC. New York and Basel

Library of Congress Cataloging in Publication Data

McCord, Bruce E., [date]
   Designing pneumatic control circuits.

   (Fluid power and control)
   Includes index.
   1. Pneumatic control. I. Title. II. Series.
TJ219.M33   1983   629.8'045   83-2103
ISBN 0-8247-1910-7

Copyright © 1983 by Marcel Dekker, Inc.   All Rights Reserved

Neither this book nor any part may be reproduced or transmitted in any form or by any means, electronic or mechanical, including photocopying, microfilming, and recording, or by any information storage and retrieval system, without permission in writing from the publisher.

MARCEL DEKKER, INC.
270 Madison Avenue, New York, New York 10016

Current printing (last digit):
10 9 8 7 6 5 4 3 2 1

Printed in the United States of America

ML

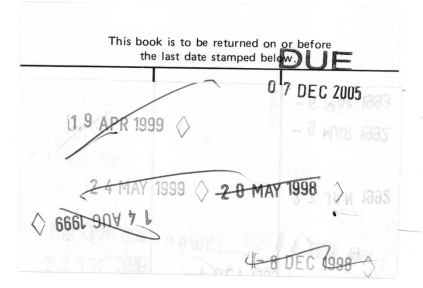

DESIGNING PNEUMATIC CONTROL CIRCUITS

FLUID POWER AND CONTROL

A Series of Textbooks and Reference Books

Consulting Editor

Z. J. Lansky
Parker Hannifin Corporation
Cleveland, Ohio

Associate Editor

F. Yeaple
TEF Engineering
Allendale, New Jersey

1. Hydraulic Pumps and Motors: Selection and Application for Hydraulic Power Control Systems, *by Raymond P. Lambeck*
2. Designing Pneumatic Control Circuits: Efficient Techniques for Practical Application, *by Bruce E. McCord*

Other Volumes in Preparation

Hydraulic Valves and Controls: Selection and Application, *by John J. Pippenger*

Fluid Power Troubleshooting, *by Anton H. Hehn*

To my father
Harry J. McCord
who put up with my very extended learning curve.

You were right, Pop.

PREFACE

This book is intended to explain moving part pneumatic controls to anyone, regardless of their previous experience in controls. It explains the basic principles and advances very rapidly through the individual functions into the design of complete circuits.

To accomplish this, one design method has been selected and is described in detail. This method works equally well with any type of moving part pneumatic control hardware, be it spool, poppet, diaphragm, or combinations of these.

The symbols used throughout this book are from the American National Standard Method of Diagramming for Moving Part Fluid Controls (ANSI B93.38-1976-R1981), sponsored by the National Fluid Power Association, Inc. (NFPA Std. T3.28.9-1973). The discussion of the attached method includes proposed additions. A cross-reference to the detached method is also included.

My sincerest appreciation to Beth Zuver and Tom Burton for their efforts in preparing the manuscript and illustrations for this book.

Bruce E. McCord

CONTENTS

Preface   v

1. **History and Basic Principles**   1

   History / Today's Uses / The Future / Basic Principles

2. **Manual Input Devices**   11

   Introduction / Push-Button Valves / Selector Valves / Designing with Push Buttons and Selectors / Indicators

3. **Mechanical Input Devices**   17

   Limit Valves / Two-Way Valves / Pressure Build-Up or Decay / Noncontact Sensing

4. **True Logic Elements Or-And-Not**   29

   Introduction / True Logic Elements

5. **Timing Functions**   44

   Basic Principles / Timer, Delay, and Pulse Elements / Basic Timing Functions

Contents

6. Memory Functions — 49

Basic Principles / Variations in Memory Functions / Memory Circuits Using True Logic Elements / Special Memory Elements / Common Circuits Using Memory and Delay Functions

7. Special-Purpose Elements — 70

Basic Principles / Electrical to Air Interface / Air to Electrical Interface / High to Low Pressure / Low to High Pressure / Flow Reduction and Increase / Three-Way to Two-Way Interface / Two-Way to Three-Way Interface / Air to Hydraulic / Hydraulic to Air / Air to Vacuum / Vacuum to Air

8. Sequential Circuit Design Basic Principles — 79

Introduction / Sequential Circuit Design Method / Modifications to Basic Design

9. Variations of Sequential Circuits — 111

Introduction / Expansion / Mixing / Contraction / Double Sequence / Continuous Operation / Driven Sequence / Electrical Comparison / Repeaters

10. Input Variations and Other Special Circumstances — 125

Introduction / Bleed Functions / Pressure Sensing (Increasing) / Pressure Sensing (Decreasing) / Other Input Modifications / Timing

11. Additional Circuits and Approaches — 148

Introduction / Start/Stop Circuits / Liquid Level Sensing / Master/Subcircuits / Test Circuits / Conclusion

Appendix A: Input Symbols — 159
Appendix B: Circuit Symbols — 161
Appendix C: Output Symbols — 163

Index — 165

1
HISTORY AND BASIC PRINCIPLES

HISTORY

While the use of compressed air is by no means new, the use of compressed air as a control has only recently become practical. The basic principles have been well understood for years, but the lack of true control hardware limited the application of pneumatic control to only the most extreme cases, such as explosive environments.

Early pneumatic control circuits were designed using three-way and four-way "power valves." These power valves were themselves designed to extend and retract air cylinders or operate air motors and other power devices. They were not necessarily designed to work well with each other and create a control circuit. To design these systems, the designer needed to be totally familiar with all of the idiosyncrasies of the valves, plus have the ability to design a circuit. The installation of the circuit was also a problem since it involved large and rather awkward valves connected together by hose or steel tubing. Needless to say, practically everything having to do with control was much more easily accomplished with electrical relays, at a much lower cost.

In the late 1940s and early 1950s, two things happened which made the use of pneumatic controls somewhat more attractive. The first was the availability of small-diameter plastic tubing and fittings. The second was the miniaturization of air cylinders and valves. These small manual, mechanical, and pilot-operated valves, although still designed as power valves for small cylinders, greatly reduced the space required to install a system. They also reduced the cost and the amount of labor required to install a

2 History and Basic Principles

system. Air control systems then became practical in an increasing number of applications. Still, the design of circuits using these valves required a thorough understanding of the way in which they would interact with each other, *plus* the ability to design the circuit. Seldom could you expect the controls to function exactly as designed.

The circuits were also difficult to install. Each component had to be mounted individually, and if not carefully planned, the system could easily become a nightmare of tubing and fittings. In spite of this, pneumatic controls found greater use in several industries because of individuals who became expert at applying air controls to specific types of equipment.

Finally, in the mid-1960s, several brands of pneumatic controls were introduced. These moving-part logic systems were the first air valves specifically designed to be used as air controls. They were still three-way spool or poppet valves; however, the considerations given to various aspects of their design were quite different from those of a power valve. Here are some of them.

Physical size: In designing an air control system, smaller is better (with some limitations). All of the components should be bottom ported to reduce tubing clutter. The elements should be uniform in size and shape so that they fit together compactly regardless of the mix of components used. They must be easy to mount and interconnect into a package. In fact, anything that can be done to make the controls easier to assemble and neater (less tubing clutter) is definite benefit to the user. Because they are used together and are quite similar in appearance, the identification of the function of each component is important, and further, a system to identify like functions by number or location is necessary. Test ports or indicators for troubleshooting are also incorporated into these components.

Function: Snap action—ability to shift quickly and positively from one position to the next—is desirable. A wide deadband, which allows the components to ignore slight variations in pressure, is also very important. The elements should be designed so that they all shift at the same pressure, increasing and decreasing, reducing potential "racing circuits." The flow must be sized to the type of connection used (tubing, fitting, or channels) for the optimum response time possible. This also includes the exhaust flow capabilities of each element, since it is also important to get rid of a signal quickly. These considerations extend beyond the control elements themselves and out to the push buttons, limit valves, and even the tubing and fittings used to connect them into the system.

Operating conditions: Here the key word may well be uniformity. Just as a chain is no stronger than its weakest link, a control system will be no better than its weakest component. In general, all components must be nonlube since it will be impractical to evenly distribute lubrication into the control system (besides being messy when it exhausts to atmosphere). They must all operate over a wide range of pressure, or they must all operate at one ideal pressure with design provisions to maintain that pressure. They must all operate on a certain "quality" of air. Tolerable levels of dirt and moisture are established. This, then, determines what type of air line filters, for example, are required with each control.

Life expectancy: Any weakness here would also subject practically all systems to premature failure. Minimum design standards of 50 to 100 million cycles are common in pneumatic controls. Further improvements with field experience have made these devices nearly indestructable for all practical purposes.

In reviewing the efforts made to meet these specifications, it is apparent that the brands of controls introduced have both recognized and accomplished most, if not all, of these objectives. This has made the design, installation, and use of pneumatic control systems much easier and less expensive than ever before. Again, the number of applications in which pneumatic controls could be justified increased.

An educational cycle began. The control hardware at this point was probably better than our ability to use it. The design of control circuits had been, up to that time, almost an art form performed by those experts who developed many different methods of design peculiar to their industry and the type of hardware they used. For the use of air controls to truly expand, two additional steps were required.

The first was a common language which could be understood by all who were involved in the design and maintenance of air control systems. This language consists of the symbols used to represent the various control components on a drawing. A method did exist; however, it was developed primarily to represent power valves. This standard (American National Standard ANSI Y32.10), although great for diagramming power valves, was very difficult to use when designing or troubleshooting control circuits. It has been said that designing control circuits with this standard was similar to figuring your income tax using Roman numerals. Most of the pneumatic control experts had already developed their own shorthand methods of describing control circuits, but of course, they were all unique.

So, at the request of industry, the National Fluid Power Association (NFPA) called a general conference on July 20, 1967 to explore the need for a standard method of diagramming for control circuits using moving-part fluid controls. It was agreed that the need did exist and that the NFPA should sponsor this activity.

Early in its work, the project group decided to offer two methods of diagramming moving-part controls. The *detached* method is similar to an electrical relay ladder diagram. This method separates the pilot port of the valve from its flow path. It can also show different flow paths of the same valve at different points in the diagram. This method had proven to be the best method to describe multipurpose valves (normally, spool valves) when used as a control.

The *attached* method is similar to the diagramming method used for electronic controls. This method shows the complete symbol for each component with all of the input and output symbols connected directly to the symbol. This method had proven to be best in describing single-purpose valves (normally, poppet valves) when used as a control.

Standard symbols for input devices (push buttons, selector valves, limit valves), and output devices (such as power valves, and cylinders) were also developed. These symbols are the same for both the attached and detached diagrams and also borrow heavily from electrical standards.

On July 19, 1973, the final document was approved as a recommended standard by the NFPA Board of Directors (NFPA T3. 28.9-1973). On March 4, 1976, the document was granted approval by the ANSI Board of Standards Review and became ANSI B93.38-1976. In 1981 this standard came up for 5-year review and was reapproved.

On March 25, 1982, the USA Technical Advisory Group (USA TAG/SC5) voted to submit the ANSI B93.38 as a U.S. proposal for the ISO (International Standards Organization) standard on moving-part logic controls. So, gradually an industry-wide (national and, it is hoped, international) standard language used to communicate air control concepts has been developed.

The second step is training—developing material that will explain the basic concepts, the language, and the techniques involved in designing pneumatic controls. This book is a part of that effort. Several books have been written which describe the various types of hardware now available. Several others give brief explanations of several methods of designing pneumatic controls. What we want to do here is to give a very detailed, in-depth description of the language of controls and a method of designing circuits with as many examples as possible in one book. In order to do this we have decided to do the following:

1. Refrain from including product pictures, cutaways, and other product information.
2. Use only the attached method of diagramming (with a cross-reference to the detached method in the appendix).
3. Describe one method of circuit design as completely as possible. The method selected for sequential controls can be used with any type of hardware or diagramming method. It has been prov-

en in hundreds of applications of many types and is quite easy to understand and use. In addition, it differs very little from other methods in its basic concept.

TODAY'S USES

Today, air controls are used in a very wide variety of applications. Understanding where and why these air controls are used is important. The graph in Figure 1 attempts to show the types of machines, in standard environments, for which pneumatic controls might be justified. This graph is intended to show all machines, from the simplest fixture to the most complex of robots and systems (left to right). The graph also shows them divided into three groups by the type of power used on the machine. Obviously there is some mixing here, but for the purposes of this graph it is being ignored. An arbitrary line is drawn on this graph to show what machines would be likely candidates for pneumatic controls and those which are more likely to be electrical. This is an economic consideration.

The single largest reason for the use of air controls is the reduced cost of the system. The cost savings generated by the use of pneumatic controls

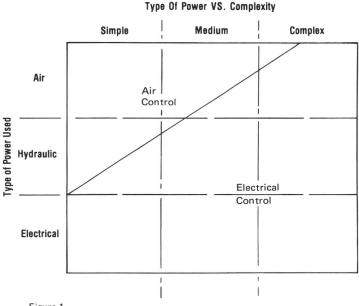

Figure 1

6 History and Basic Principles

is not always readily apparent. Quite often the savings is generated on items not a part of but affected by the control circuit. A simple example of this cost savings is shown in Figure 2. This application involves an air cylinder, a four-way power valve, and a foot valve or switch, depending on whether air or electrical is used as the controlling medium. To use electrical controls, you must make provisions for the electrical power to this machine. This will include conduit, wire connections, and fusing, among other things, plus labor to install it.

Compare this to the cost of using the penumatic pressure already necessary on this machine (to operate the air cylinder) and you can see the first cost savings. To use air as the control medium on this machine, you simply tee from the air inlet down to the supply of the foot pedal valve. The second savings is the connection from the foot pedal to the power valve. This is either a tube and two fittings or wiring, conduit, conduit connectors, and any other material required plus labor. The third savings is in the power valve itself. The cost of a single-solenoid, spring return four-way air valve is greater than the same valve in an air pilot operated version (by approximately $35 to $50 at the time of writing). Since the cost of the foot valve versus the foot switch is about the same and the balance of the components is identical, the cost savings in using air as the control for this machine is obvious. Future savings in reliability and the fact that a single trade is required for any maintenance could also be considered a plus on this selection.

In spite of this, it is estimated that less than 20 percent of the machines that are identical to the one just described are controlled by air in the United States. Habits, developed during the time when air controls

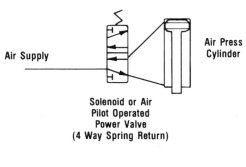

Solenoid or Air
Pilot Operated
Power Valve
(4 Way Spring Return)

Foot Operated Electrical Switch
or Air Valve

Figure 2

were not readily available, can be the only reasonable explanation for this.

This application would be in the uppermost left-hand corner of our graph in Figure 1. This would also be considered the hard-core market for pneumatic controls due to its extreme cost justification and the fact that it is an old market (the pneumatic hardware needed for this application has been available for years). Justified air control applications extend from this point to the right (to more complex machines), and some extend down (into hydraulic-powered machinery). Just how far to the right they can be justified is dependent on several factors.

The first is the input/output to memory ratio. Very simply, this means the amount of control required (which is more expensive) versus the number of inputs and outputs (which is a savings). For example, a machine that has 100 logic functions and 6 inputs and outputs could require an electrical control. However, a machine, again with 100 logic functions but with 20 inputs and outputs, could more economically be designed with air controls.

The second is environment. Explosive atmospheres, corrosive atmospheres, washdown, dust, dirt, and other conditions can extend the line of justification into even more complex control packages than would normally be considered for air controls. These factors magnify the cost of applying electrical controls, but often add little if any additional cost to the pneumatic control system.

The cost justification drops off when applied to machines which use hydraulic power. That is because first, there is no evidence that an air supply exists on this machine and second, an electrical connection is often required to operate the hydraulic pump. Other than this the cost advantage of using air (such as pilot valves versus solenoids) remains.

A further reduction in the cost justification occurs when air controls are used on machines that employ electric motors and other electrical devices. In this type of machinery, the interface cost advantage is reversed. The pneumatic signals must be converted (normally through the use of a pressure switch) before they can command the electrical power devices.

So, as the graph illustrates, the areas where pneumatic controls can most often be used to advantage are air-powered machines and processes in the simple to medium complex range, with some applications in hydraulic-powered machines as well.

THE FUTURE

The need for control systems of all kinds is great and is expected to continue to increase in the future as the pressures to automate and increase productivity are felt. The need to automate is not limited to robots, huge systems, and programmable controls, but extends to practically every type

8   History and Basic Princples

of machine and system, large and small, simple and complex. Many of these can best be controlled by pneumatic controls. The number that actually will be controlled with pneumatic controls will be related much more to the number of people capable of designing these controls than to any other single factor. The designer must first become aware of the potential and then be able to design reliable, safe, and economical systems in order to take advantage of this technology.

New developments in air controls continue which make them more versatile, more reliable, easier to use, and less expensive. Air controls are not a panacea, but are tools which, when properly selected for use, then used properly, can be a tremendous benefit to the designer. It is expected that the uses of air controls will increase in the future as more and more designers become aware of the benefits and knowledgeable in their application.

BASIC PRINCIPLES

The basic devices used in pneumatic control are three-way valves. These valves have two positions. In one position they connect the output port to pressure (from supply). In the second position they connect the output port to atmosphere (Figure 3). Even though this is a three-way valve that closes one passageway as it opens another, it creates only two conditions at the output port. These conditions are:

1. The output is pressurized, called ON or 1.
2. The output is not pressurized, called OFF or 0.

Because air controls of this type have only two output conditions, they are called binary devices. This term clearly separates these controls from other types which have variable output pressures and are called

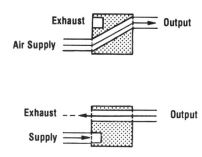

Figure 3

Basic Principles 9

analog devices. So, in studying this material, you must consider only two possible signal conditions—ON and OFF.

In the ON condition, a channel or series of channels is open all the way back to the pressure source, normally a compressor. Since all of the control devices are supplied through one regulator, the ON pressure that exists at any point in a system should be equal to that at any other point. There is no continuous flow; lines and valves are merely filled or discharged, and in the ON condition small leaks do not affect the system (because they will be made up by additional air from the compressor).

In the OFF condition, this same line will be connected to atmosphere. The pressure in this line will be 0, and any leakage that might occur into this line will immediately be discharged to atmosphere.

There are several advantages to working with an air control system. Some of these advantages are as follows. Because there are only two possible output conditions, the design of control systems is simplified. Faults in the system are quite obvious since they normally result in a steady flow of air to atmosphere. The flow capacity and pressures used prohibit small leaks from affecting the function of a circuit. Also, because they are air controls, they are very forgiving. Design or installation mistakes can be made without damage to the components themselves.

Air control systems are made up of four basic types of components. They are:

1. Operator control devices: These consist of push-button valves, selector valves, and indicators.
2. Operation sensing devices: Limit valves, noncontact sensing devices, and other devices to sense the operation of the machine or process make up this group.
3. Circuit hardware: These are the logic elements of various types which make up the control circuit itself.
4. Power devices: Cylinders, valves, motors, and other devices that control and perform the work being done by machines and processes.

This book will cover the first three of the four groups in the order listed. These are the operator control devices, operation sensing devices, and circuit hardware.

The power devices are a complete subject in themselves. Although we describe some of them briefly in the section devoted to interfacing, further study of other works on the subject of power devices, such as pneumatic and hydraulic cylinders, motors, and valves, may be necessary to completely prepare yourself for the design of a machine or system.

10   History and Basic Principles

The three groups that will be discussed will be described in the following manner: First, the overall purpose of the group, and second, how it will be identified on a drawing. For this we will use the symbols contained in the American National Standard Method of Diagramming for Moving Part Fluid Controls (ANSI B93.38-1976-R 1981). As of this writing, the Standard is under revision. This revision will add additional symbols to the document. The additional new symbols are also used in this book. Finally, application tips and examples of their use will also be included.

The final chapters of this book will be devoted to the actual design of several circuits, including methods of design that have proven successful on many actual applications.

2
MANUAL INPUT DEVICES

INTRODUCTION

Push-button valves, selector valves, and pneumatic indicators are operator control devices. Push-button valves and selector valves allow the operator to communicate with the machine control system. They convert a manual action, such as pushing a button or turning a lever, into air signals compatible with the air control systems—signals that the air control systems can recognize and respond to. Indicators perform the opposite function. They convert air signals into visual indications that the operator can understand and respond to.

In many cases, the operator becomes part of a control "loop." The operator sees certain conditions, either directly on the machine or with the aid of indicators, and then responds to these indications by issuing commands back to the machine (by pushing buttons or moving selector valves).

PUSH-BUTTON VALVES

The push-button valve symbol is made up of several parts. The first part is the terminal. (See Figure 1.) The terminal is a universal symbol used on practically all input devices for an air control. The open circles on the terminal show that this is a three-way valve, capable of the two positions previously described.

1. Connecting pressure to the output from the supply port
2. Blocking this supply and connecting the output to exhaust (atmosphere)

12   Manual Input Devices

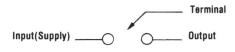

Figure 1

Thus, the terminal to the left represents all of the functions of the valve symbol to the right. (See Figure 2).

Added to the terminal are symbols which represent a push button, called a bridge (Figure 3). And in cases in which the push button is returned by a spring, this is also shown in the symbol (Figure 4). The total symbol thus represents a three-way, normally nonpassing valve, operated by a push button and, when released, returned to its "normal" (nonpassing) position by a spring (Figure 5).

The two drawings in Figure 5 represent the same push-button valve. If you have had some previous experience in air circuits using the older style of valve symbols on the right, you may at first feel more comfortable with this type of diagram. However, if you take the time to learn and work with the newer symbols recommended for control circuits, you will soon find that they are much easier to draw and their function is much easier to recognize at a glance. For example, the two valves in Figure 6 have entirely different functions. Yet at a glance they look almost the same. It is only when you look closely at the connection point of the supply line and the direction of the arrows that you realize that the one on the left is a normally passing, spring return push-button valve and that the one on the right is a normally nonpassing, spring return push-button valve.

These same push-button valves are shown in Figure 7 using the control symbols. As you can see, the difference in their function is very obvious since the bridge on the left is held against the terminals by the spring (normally passing) and released by the push button. The symbol on the right shows the bridge held away from the terminals (nonpassing) until the push button is actuated.

Many other types of push-button operators are available, such as mushroom head, push-pull, latching, key lock, and pilot return. These in turn are mounted to valves which can be two-way (with no exhaust port), three-way selector (with two supply pressures), and four-way action (some-

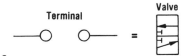

Figure 2

Push-Button Valves 13

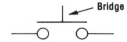

Figure 3

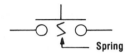

Figure 4

Figure 5

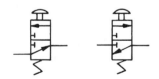

Figure 6

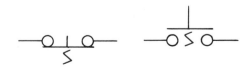

Figure 7

14   Manual Input Devices

times using two 3-way valves). For these special symbols, you will need to consult ANSI Standard B93.38-1976 (R1981).

SELECTOR VALVES

Selector valves use the basic terminal plus a bridge that illustrates the selector lever. Figure 8 shows a drawing of a simple on-off three-way selector valve. Remember that three-way refers to the valve and its function, not the lever. The solid line to the arrow and the solid line of the bridge show the condition of the valve in one position. The broken line and bridge show the condition of the valve when the lever is moved to the second position.

Often, selectors will operate more than one valve. This allows the selector to send a signal corresponding to each of its positions. Such a selector valve is shown in Figure 9. Remember that each terminal represents a three-way valve with its own exhaust port. Here we have a *two-position* selector operating two 3-way valves. In the manual position shown by the solid line and arrow, the upper three-way valve is passing (manual output on). When the lever is selected to "auto," as indicated by the broken lines, the manual output will go off and the automatic output will go on.

Air selector valves are available with actuators for three and more positions, and rotary selector valves often have individual valves for each one of these positions. In addition, special operators are available, such as key-operated spring return and spring centered. For these special symbols, you will need to consult ANSI Standard B93.38-1976 (R1981).

DESIGNING WITH PUSH BUTTONS AND SELECTORS

The variety and complexity of push buttons and selectors can be a point of confusion to the designer, especially in the early stages of the circuit design. By following a few simple rules, you can eliminate much of this confusion. First, in the early stages of design, use only single valves on spring return operators for push buttons and single valves on two-position maintained operators for selectors. By doing this, you will be dealing with only the two symbols shown in Figure 10.

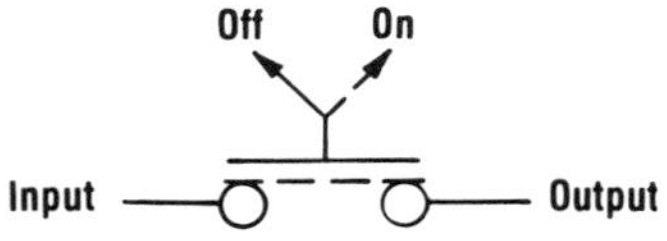

Figure 8

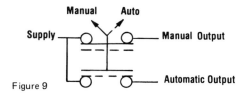

Figure 9

Figure 10

Second, you will be dealing with only two types of input signals. The signal from the push button will be momentary (on only as long as the operator holds the button). The signal from the selector valve will be maintained.

Once the basic circuit has been designed using these two simple devices, the push buttons and selectors can be reviewed and combined or changed into more complex functions. This, then, can be done with an eye to operator convenience, reduced number of components, and cost.

INDICATORS

Indicators are to pneumatic circuits what indicator "lights" are to electrical circuits. They are used to advise or warn the operator of various conditions of the machine using color change as an indication.

Indicators are also used as repair tools. Strategically placed indicators can be arranged to indicate the probable cause of potential trouble in a machine or process. These indicators can then become a means of diagnosis when problems develop.

Visual indicators (pressure-operated spring return) are shown in Figure 11. The color of the indicator when pressurized is normally noted in the center of the indicator symbol. The indicator legend, or what it tells the operator, is also noted beside or above this symbol.

Two-position detented indicators are also available. These are sometimes used to monitor momentary pressure signals. A momentary signal at port A causes the indicator to change color. When the B port is pressurized, the indicator returns to its starting condition (Figure 12).

16   Manual Input Devices

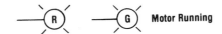 Motor Running

Figure 11

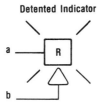

Figure 12

3
MECHANICAL INPUT DEVICES

LIMIT VALVES

Operation sensing devices detect machine functions and speak directly to the control system. The most common of these is the limit valve. Again, it is a three-way valve, but in this case it is operated mechanically by a motion of the machine itself. The operator does not enter into this action.

The limit valve shown in Figure 1 is an operation sensing device. It has been assigned the responsibility of detecting when the part is clamped.

When the air cylinder extends and clamps the part, it also actuates the lever on the limit valve. When the limit valve is actuated, a pressure signal is sent to the control circuit. This pressure signal is interpreted by the control circuit, because of its design, to mean that the part is now clamped. Thus, this valve output indicates two conditions to the control circuit. OFF means that the part is not clamped. ON means that the part is clamped. Again, this is the binary (two-part) function of a three-way valve.

Designing with Limit Valves

The limit valve drawing uses the basic terminal to indicate a three-way valve function. (See Figure 2.) To this terminal is added a bridge which symbolizes a limit valve. (See Figure 3.)

How this bridge is placed on the terminal will show: (1) how the valve is connected (normally passing or normally nonpassing), and (2) whether or not this limit valve is actuated at the beginning of the cycle.

18 Mechanical Input Devices

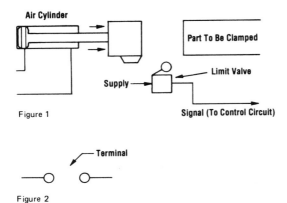

Figure 1

Figure 2

Limit valves and other operation sensing devices are shown on the drawing in Figure 4 in the position they will assume when the machine is idle. For example, in the case of the part clamped limit valve used in the first illustration, let's say that in the idle position of this machine the part is not clamped. The clamp cylinder is retracted. The drawing for this limit valve would then show that it is a three-way, normally nonpassing valve, not actuated at the beginning of the cycle. When the initial position of the limit valve cannot be determined, the valve should be shown not actuated.

Since there are no springs shown in limit valve symbols, it will be helpful for you to remember two rules. Rule 1 is a force which approaches this valve symbol from the top of the drawing. You might consider this gravity, since it is a constant. Our example of a part clamped limit valve has only one force acting on it in the idle condition (Figure 5).

Rule 2 is a force applied to the pad on the bridge symbol. This is the actuating force against the operator of the limit valve. If the valve were actuated at the beginning of the cycle, the actuating force (two) would overcome force one and the valve would be shown in its actuated position (Figure 6). Thus, the valve would be drawn with the bridge touching both terminals. This valve is still connected normally nonpassing; however, it is shown in its actuated position because this is the position it assumes at the end of the cycle.

To show a valve piped normally passing, the bridge is shown on top of the terminals (Figure 7). Now you can see that with no force on the operator, the valve will be in the passing condition (terminals connected).

The symbol for a normally passing valve, actuated at the beginning of the cycle, is shown in Figure 8.

Limit Valves   19

Bridge For Limit Valve Function

Figure 3

Part Clamped

Figure 4

Force 1

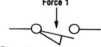

Figure 5

Force 2
Figure 6

Force 1

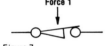

Figure 7

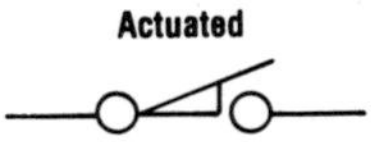

Actuated

Figure 8

Released

Actuated

Released   **Actuated**

Figure 9

20 Mechanical Input Devices

The four symbols shown in Figure 9 show all the possible conditions of a three-way limit valve.

TWO-WAY VALVES

Two-way valves, also called bleeder valves, are sometimes used to replace three-way limit valves. The function of a two-way valve is only to trap or exhaust a pressure signal supplied from another source. The simplicity of its function allows for design of devices that have special features. The two-way valve shown in Figure 10 (on the left) is very compact and easy to install. The drawing on the right shows how this two-way valve could be installed and connected by merely drilling and tapping through the fixture. Only one connection is required since the supply is generated from another source. In this instance, a two-way valve may be used because it is more compact and easier to install than a three-way valve.

In other cases, the two-way valve can be designed to require lower forces to operate. "Whisker" valves are sometimes used to detect very light objects or motions because they can be designed to operate by forces of one-quarter ounce or less (Figure 11).

Designing with Two-Way Valves

The symbol used to show a two-way valve in a circuit is very similar to a three-way limit valve. The circles on the terminal of the valve are filled in to denote a two-way valve as shown in Figure 12. The dash lines indicate a port which vents to atmosphere.

The bridge is added to the terminal using the same rules as the three-way limit valve. The drawing in Figure 13 shows a two-way normally non-passing valve which is not actuated at the beginning of the cycle.

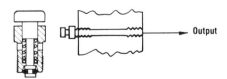

Figure 10

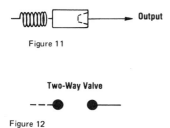

Figure 11

Two-Way Valve

Figure 12

Two-way valves are not directly compatible with most moving-part air logic controls. To be used in a system, the signal from this device must be converted to a "three-way" (pressure/exhaust) signal. This function is accomplished by a special device called a two- to three-way converter element. A drawing of the complete function including the two-way valve and the converter element is shown in Figure 14. Additional information on the function of the two- to three-way converter element is given in Chapter 7.

PRESSURE BUILD-UP OR DECAY

Another method of sensing machine actions is with pressure build-up or decay. Figure 15 shows a double-acting air cylinder, a four-way valve, two flow control valves, and two gauges. The cylinder is shown in the retracted position. There is no pressure on the blind end of the cylinder, but there is pressure on the rod end holding the piston retracted.

The charts in Figure 16 show what will happen to the pressure on each gauge when the valve shifts, the cylinder extends, and the cylinder stops.

By connecting a pressure-sensing element in place of gauge A, the movement of the cylinder can be detected by sensing the increase in pressure. Notice that the pressure does not approach full line pressure until the cylinder has completed its movement. In this case, the pressure-sensing element would be adjusted to trip (shift) on an increasing pressure near maximum. This is shown as point A on the increasing chart.

The movement of this cylinder can also be detected by sensing the decline in pressure at the opposite port. In this case, a pressure-sensing element would be connected in place of gauge B. Notice again that the pressure in the rod end of the cylinder does not go to zero until the cylinder has

22 Mechanical Input Devices

Figure 13

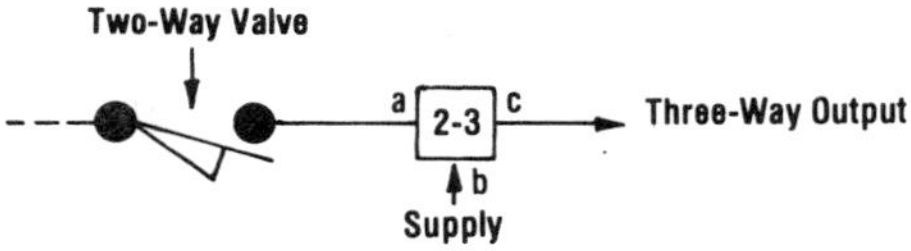

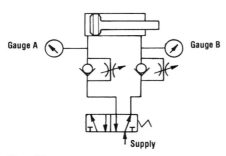

Figure 14

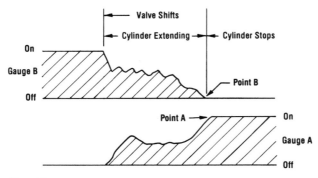

Figure 15

Figure 16

Noncontact Sensing 23

completed its movement. In this case, the pressure-sensing element would be adjusted to trip at a decreasing pressure near zero. This is shown as point B on the pressure chart. This method offers some additional advantages. In many cases, these devices need not be adjustable since zero is an absolute, whereas a pressure-sensing device to sense increasing pressures will, by its nature, require an adjustable feature.

Designing with Pressure-Sensing Elements

There are some occasions when the use of pressure to sense the movement of a cylinder is very desirable. An example of an application of this type is shown in Figure 17. The cylinder shown here is used to clamp variously sized boxes. In doing this, the cylinder will stop at various positions as it extends, depending on whether it is clamping box A, B, or C. In this application, pressure sensing could easily be the best alternative to sense that the cylinder was holding the box and start the next step of the sequence.

There are other cases in which pressure sensing can be used, such as in cases where a limit valve would be difficult to locate. In these instances, as well as in the application described previously, the designer should be very conscious of one thing. If the cylinder is jammed for some reason and fails to completely extend, the signal from the pressure-sensing device will nevertheless occur. Thus, the control system will "think" the cylinder has extended. This fact must be considered each time pressure sensing is used.

The symbols for the pressure-sensing elements are below and further described in Chapter 7. (See Figure 18.)

NONCONTACT SENSING (AIR JET SENSING)

Noncontact sensing is sometimes required for applications involving low force (such as light products), products that cannot be touched for other reasons (such as freshly painted parts), and for very accurate sensing and positioning.

The simplest form of noncontact sensing is an air jet. An illustration of an air jet system is shown in Figure 19. The air flowing through the regulator and adjustable orifice is reduced in pressure and volume to the point

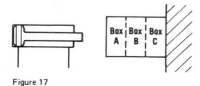

Figure 17

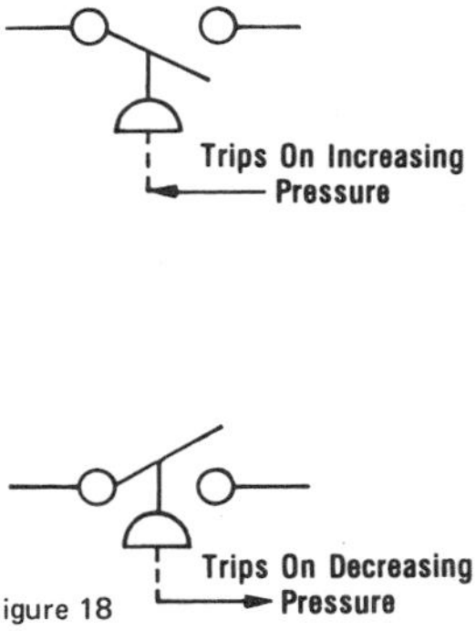

24 Mechanical Input Devices

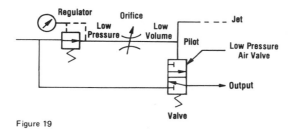

Figure 18

Figure 19

that it cannot overcome the spring on the valve with the jet flowing freely. When the jet is restricted by moving a part close to it, a back pressure builds up in the line and shifts the valve.

The rule of thumb for this type of sensing is that you can create a back pressure when the part is at a distance of approximately one-half the diameter of the orifice. From that point on, it depends on the sensitivity of the valve. This means that for all practical purposes, you must *block* the jet completely before you get a response from the valve. Although this type of jet sensing is in many cases impractical, there are occasions when it can be used. For example, because the jet is *not* sensitive, it also becomes very accurate. In other words, the item being sensed must be in the exact position called for before the signal will be given. The symbol used to show a jet sensor of this type is shown in Figure 20.

You will notice that in jet sensing there also is a means of showing whether the jet is blocked at the beginning of the cycle. This is done by the position of the target. The jet in Figure 20 is shown not blocked. The jet in Figure 21 is shown blocked at the beginning of the cycle.

Because of the problems involved in straight jet sensing, special hardware has been developed to improve this future. The first of these is a

Noncontact Sensing 25

very sensitive valve called an amplifier. This valve will shift with a very low pilot signal, down to one-half of one psi. By using an amplifier in this circuit, the regulator can be adjusted down much lower, and the problem of a large-volume high-pressure air loss is partially corrected.

The symbol shown in Figure 22 is the symbol used for an amplifier element. Additional information on this element is given in Chapter 8.

By developing special hardware to replace the jet, parts can be sensed at greater distances. Although many types of hardware are available, they can be classified as two basic types. The first is called a noncontact position sensor. (See Figure 23.) Internally, this jet is constructed as shown in Figure 24.

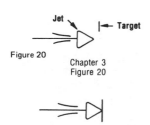

Figure 20

Chapter 3
Figure 20

Figure 21

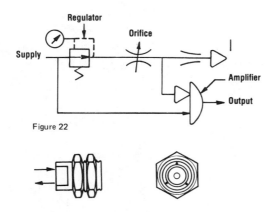

Figure 22

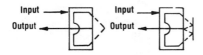

Figure 23

Figure 24

26   Mechanical Input Devices

Figure 25

Figure 26

The input signal (low pressure) is divided into three jets or more (only two are shown here) and sent from the face of the sensor at an angle. These three jets form an "air cone" beyond the face of the sensor. This air flow causes a slight negative pressure on the output (vacuum). When an object approaches the face of the sensor, the cone is broken and a small amount of air flow is directed back toward the output connection. This causes a slight positive pressure which then shifts the amplifier element. The symbol for this device used in a circuit is shown in Figure 25.

The second sensing nozzle is called an interruptable gap sensor. (See Figure 26.) The internal function of the interruptable gap sensor is shown in Figure 27.

The interruptable gap sensor, when not interrupted by the target, is producing an output signal. Air from the input flows out jet 1. It also flows through an orifice which feeds a lower volume of air to jet 2 and the output. The force of the air flowing out of jet 1 restricts the flow out of jet 2 and creates a back pressure on the output. When an object enters the gap (see Figure 28), it interrupts the flow from jet 1 to jet 2. Jet 2 can then flow freely to atmosphere—faster than it can be refilled through the orifice. The pressure at the output port then drops to zero.

The symbol for an interruptable gap sensor when used on a circuit diagram is shown in Figure 29.

Notice two things about these sensor devices. First, both are self-cleaning because the negative signal created on the noncontact position sensor draws material out of this orifice. Both ports are designed to flow outward on the interruptable gap sensor to prevent contamination of the sensor nozzle. Second, the output signal conditions created by a part in sensing position are opposite on these two sensors. With the noncontact position sensor, you have a pressure signal with the part in place. With the interruptable gap sensor, you have *no* signal with the part in place.

These signals can be reversed by the use of different types of (normally passing or normally nonpassing) amplifiers later in the system. These devices, as well as the complete drawings for sensing systems, are shown in Chapter 8.

The size of part that can be detected using an interruptable gap sensor is limited to the size of the opening on this sensor. (See Figure 30.) To sense larger gaps and larger parts, an emitter is added to this circuit. (See Figure 31.) The emitter is nothing more than an additional jet positioned to interrupt the flow from jet 1 to jet 2. The two devices are diagrammed in Figure 31 without the target in place. In this condition, there will be no output from the gap sensor.

When a target is in position, the air flow from jet 3 to the gap sensor is stopped. The air flow from jet 1 to jet 2 becomes uninterrupted, creating a back-pressure signal at the output of the sensor. (See Figure 32.)

Designing with Noncontact Sensors

Although the drawings and descriptions of sensing can become very complicated, the actual use of these systems can be relatively simple. Complete systems are available, completely assembled, to do these sensing jobs. These systems contain an air filter, low-pressure regulator and gauge, and the sensors and amplifiers required to reduce the shop air pressure to sensing pressure, sense the part, and return a high-pressure output signal. Noncontacting systems are, nonetheless, generally more complicated, more expensive, and more temperamental than a simple limit valve. Their use should be limited to applications in which their features are clearly required.

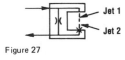

Figure 27

Figure 28

Figure 29

28 Mechanical Input Devices

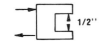

Figure 30

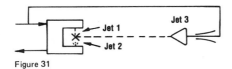

Figure 31

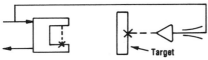

Figure 32

4
TRUE LOGIC ELEMENTS OR-AND-NOT

INTRODUCTION

Moving-part air controls have evolved from the two basic types of air valves used in industry. The two basic types of valves are spool valves and poppet valves.

Spool valves are multipurpose valves. In other words, the same basic valve can be used to perform any one of several different functions. The function of the valve is then, within its capabilities, determined by the way it is connected into the system. This is very similar to an electrical relay which could be normally passing (NC) or normally nonpassing (NO), depending on where the wires are connected. Because of this similarity of function, spool valves became known as pneumatic relays, and a system similar to an electrical relay ladder diagram was developed to show their circuits. This method is called the detached method of diagramming in ANSI Standard B93.38-1976 (R1981).

Poppet valves are generally not multipurpose. Each component is designed to perform one specific function. These functions are best described in terms of logic, such as Or, And, Not, and Memory. The similarity between the poppet valve and electronic devices has caused them to become known as pneumatic logic elements, and a system was developed, similar to the electronic method of diagramming, to show these elements in a circuit. This method of diagramming was also adapted as a standard method of diagramming in ANSI B93.38-1976 (R1981) and is called the attached method of diagramming.

Since it would be unnecessarily complicated to cross-reference both

30   True Logic Elements Or-And-Not

methods each time an example is given, this book will concentrate on the attached method of diagramming and all examples will be shown using this method. The attached method was chosen because of the ease with which the circuits and their components can be described, both with the written word and in drawings. Bear in mind, however, that the same functions can be performed by either the spool or the poppet valve and described in different terms. A cross-reference between common attached and detached symbols is shown in Appendix B.

All air circuits are nothing more than a number of elements (components) connected together to perform a specific function or functions. The first step in learning to design a control circuit is to understand the intended function of each element. These functions fall into four separate groups. These groups and the individual functions are shown in Figure 1.

TRUE LOGIC ELEMENTS

The true logic elements—Or, And, and Not—combine signals in an on-off relationship only. They do not delay or retain the signals, nor do they change them in any way. Their output will be on or off, depending only on the condition of the input signals. Because of this, they can be described using truth tables, Boolean expression, and with the written word.

The OR Element

The first such element is an Or element. The symbol for this element is shown in Figure 2. The function of the Or element is as follows: the output of the Or element will be on only if one or more of its inputs are on. Conversely, the output of the Or element will be off only if all input signals are off. By assigning the value 0 to off and 1 to on, the truth table (Figure 3) can be used to describe the function of an Or element. Notice that whenever one or more of the input signals are 1 (meaning on), the output signal is also 1.

The written expression for the Or element is A *or* B equals C. The

	All Air Controls		
(1) True Logic	(2) Timing	(3) Memory	Special (4) Function Group
1a OR 1b AND 1c NOT	2a Timer 2b Pulse 2c Delay	Memory 3a Circuits 3b S—R Memory 3c Flip-Flop	4a Amplifier 4b Pressure Sensors 4c Converters

Figure 1

True Logic Elements

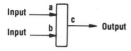

Figure 2

Inputs		Output
A	B	C
0	0	0
1	0	1
0	1	1
1	1	1

Figure 3

more complex definition says A or B or both equals C. The more complex expression is the definition of an "inclusive" Or. This means one which is also on when both inputs are on.

Figure 4 shows an Or element in a simple circuit. When push button A *or* push button B is actuated, a pressure signal is sent to the Or element. The Or element transmits this signal out the output c to shift the power valve and extend the cylinder.

This element's function in a pneumatic circuit is to prevent the air pressure from being lost out the other (nonactive) input. To explain this, you must first remember the function of the three-way valve being actuated, which is used for all input signals. In this illustration, they are push buttons.

The three-way valve has two positions. The first position connects the output of the valve to exhaust (atmosphere). The second position connects the output to pressure from supply. If an Or element were not used and the outputs were simply connected together, the pressure supplied to this connection when one valve was actuated would be lost out of the exhaust port of the unactuated valve.

The three-way output function is a characteristic of all moving-part controls, whether they are inputs as shown here or other control elements that we will describe later. For that reason, an Or element is always required when two signals are to independently control the same point.

The single Or element shown in this illustration actually becomes a two-input Or circuit. Or circuits with more than two inputs (more than two signals controlling the same output) can be made by connecting together (called cascading) several Or elements. Figure 5 illustrates a four-input Or circuit.

32 True Logic Elements Or-And-Not

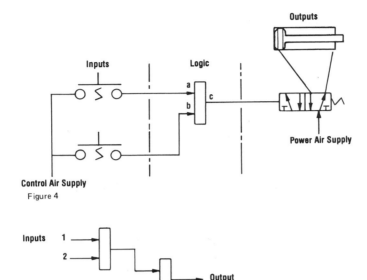

Figure 4

Figure 5

The function of the Or is also called "fan in." For all practical purposes, the number of signals and Or elements that can be combined in this manner is unlimited.

The AND Element

The symbol for an And logic element is shown in Figure 6. The function of an And element is as follows: the output of the And element will be on only if all input signals are on. Conversely, the output of the And logic element will be off if one or more of the inputs are off. Again by assigning the value 0 to off and 1 to on, the truth table in Figure 7 will describe the function of an And element.

The circuit in Figure 8 shows the And element used in a simple circuit. The cylinder on the right will not extend unless push button A *and* push button B are actuated. If either of the push buttons are released, the cylinder will immediately retract.

Additional inputs can be added to an And circuit in the same manner as the Or element. Figure 9 shows a three-input And circuit. Again, for all

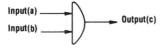

Figure 6

Inputs		Output
a	b	c
0	0	0
1	0	0
0	1	0
1	1	1

Figure 7

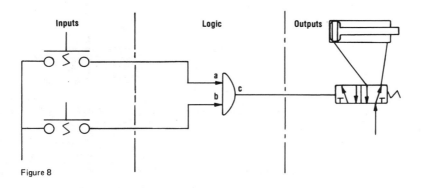

Figure 8

practical purposes, there is no known limit to the number of elements that can be used. The arrangement of And elements is of no consequence (see Figure 10). This logic function is also sometimes called a "yes" function.

Manipulation of the Elements OR and AND

Circuits which involve only true logic elements can sometimes be reduced through manipulation. Circuit manipulation follows certain rules which are based on Boolean algebra. However, as you will note from the examples shown, you need not be fluent in algebra to appreciate the various combinations that can be formed through manipulation. In fact, most manipulation can be done by examination of the logic drawing itself. Figures 11 through 14 show some examples of manipulation using the elements Or and And.

True Logic Elements Or-And-Not

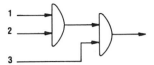

Figure 9

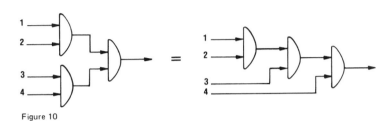

Figure 10

Rule 1: With the inputs designated by letters, the Or function is designated by a plus (+) sign between them. The Boolean expression for the logic diagram Or is shown in Figure 11.

Rule 2: The And element is represented by a dot (.) between letters as shown in Figure 12.

Rule 3: The parentheses and connectives are applied in the same manner as in ordinary algebra, even though the dot (.) and the plus (+) signs represent terms of binary logic (see Figure 13).

Rule 4: Factoring the Or and And expressions is the same as in ordinary algebra, where the dot (.) is the multiplication sign and the plus (+) is the addition sign. Factoring with ordinary algebra retains the same logic function and can be used for simplification of logic diagrams (see Figure 14).

Rule 5: The number of parentheses sets, including brackets over the whole expression, is equal to the number of logic functions. Therefore, the simplest circuit is represented by the expression with the least number of brackets.

The circuit manipulation described in Figure 14 can be accomplished in several ways. For those who prefer, the Boolean algebra method may be the most comfortable. Others may prefer to "sound it out" by reading the expression and attempting to restate it in a simpler way. Still others may prefer to draw, analyze, and redraw these circuit diagrams. The important fact to keep in mind is that this is manipulation, not design. Your first statement must be correct for any subsequent circuits or statements to be correct.

True Logic Elements   35

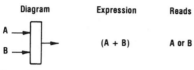

Figure 11

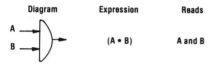

Figure 12

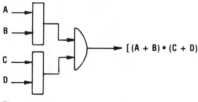

Figure 13

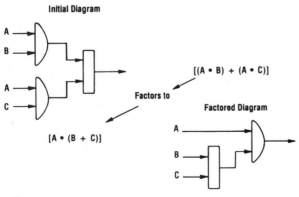

Figure 14

The NOT Element (Inversion Function)

The Not element is the final true logic element. It is always difficult to describe the function of a Not element because there are actually two functions that a Not element can perform. The two functions are the inversion function and the inhibit function. To avoid confusion, we will describe these functions separately as if they were two separate components, but bear in mind that we are actually describing the same device used in two different ways.

The symbol for a Not element used as an inverter is shown in Figure 15. When the input to the Not element is on, the output is off. Conversely, when the input to the Not element is off, the output is on. The truth table for this element is shown in Figure 16.

If you have observed that there is something missing here, you are correct. In one condition, we are getting something (output pressure) for nothing (no input pressure). This is no more true in logic than it is in real life. What is missing is a second input signal which is not shown on the drawing or mentioned in the description. This input signal is a constant pressure (supply) which is connected to each Not element, but not shown on the drawings. When the Not element is used as an inverter, this supply connection is "assumed." In other words, you know it's there, but you do not bother to draw in the signal or even mention it in the equations or descriptions of the circuit. Thus the function of the Not element when used as an inverter is shown on the chart in Figure 17.

Manipulation of the logic element Not can only be done when the Not element is used as an inverter.

Manipulation of the Elements OR, AND, and NOT

The Not function is designated with a bar over the letter or expression as shown in Figure 18.

Rule 6: DeMorgan's theorem: The Not bar can be taken off the whole

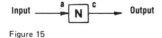

Figure 15

Not Function

a	c
1	0
0	1

Figure 16

True Logic Elements

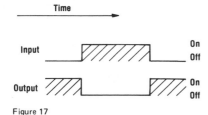

Figure 17

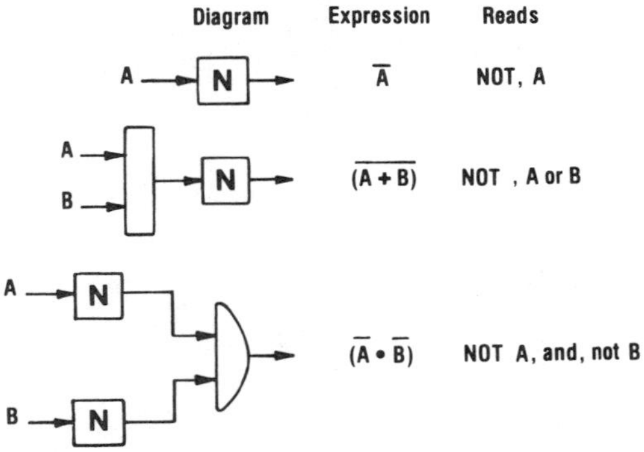

Diagram	Expression	Reads
A →[N]→	$\overline{A}$	NOT, A
A,B →[ ]→[N]→	$\overline{(A + B)}$	NOT, A or B
A →[N]→ ⊃→ ; B →[N]→	$(\overline{A} \cdot \overline{B})$	NOT A, and, not B

Figure 18

expression and placed over each term individually by changing the connectives And (•) to Or (+). (See Figure 19.)

Rule 7: DeMorgan's theorem: The Not bar can be taken off a whole expression and placed over each term individually by changing the connectives from Or (+) to And (•). (See Figure 20.)

Rule 8: Both rules 6 and 7 are reversible.

Rule 9: These manipulations apply to any number of terms (See Figure 21).

Rule 10: All of the connectives to be manipulated in one step must be either all And (•) or all Or (+). If And and Or functions are mixed in an expression, several manipulations can be used to arrive at the simplest form. (See Figures 22, 23, and 24.)

Rule 11: Equal terms with connectives Or and And in a Boolean expression are unnecessary repetitions. Examples: $(A + A) = A$; $(A \cdot A) = A$; $(A + A \cdot A) = A$. These unnecessary elements can easily be spotted either by the use of analysis of the Boolean expression or analysis of the circuit diagrams themselves (Figure 25).

Rule 12: Two Nots cancel one another (Figure 26). This rule also applies to whole expressions: $\overline{\overline{A + B + C}} = A + B + C$.

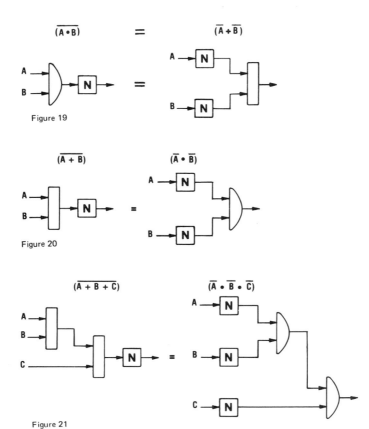

Figure 19

Figure 20

Figure 21

Initial Diagram and Expression

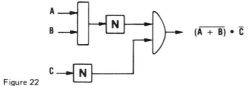

Figure 22

Second Diagram and Expression

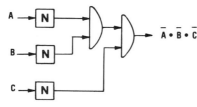

Figure 23

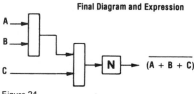

Figure 24

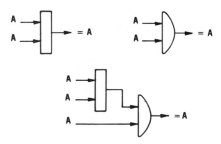

Figure 25

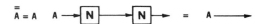

Figure 26

Figures 27, 28, and 29 are examples of circuit manipulation and redundancy. Figure 28 shows DeMorgan's theorem applied, and Figure 29 shows the redundancy theorem applied.

Remember these basic facts when using theorems:

1. Only Or, And, and Not elements can be included in Boolean expression.
2. The Not element can only be included in its inverter function.
3. This is only manipulation, not design. Your first expression must be correct.
4. Be practical—you should be able to draw and analyze a circuit at any stage of the manipulation.

The NOT Element (Inhibit Function)

We have previously stated that the Not element actually has two inputs. An inhibit function is performed when both of these inputs are used as variables. In other words, rather than just assuming that the second input is a maintained supply, the inhibit function recognizes that this signal

40   True Logic Elements Or-And-Not

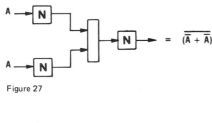

Figure 27

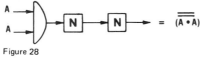

Figure 28

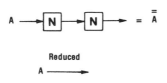

Figure 29

could also be either on or off. The Not element then becomes a true two-input element. The symbol for a Not element in an inhibitor role is shown in Figure 30. The truth table for the Not element used as an inhibitor now becomes more complex (Figure 31).

To put this into perspective, the inhibit function of the Not element is much more commonly used in actual circuit design. The use of the second input can reduce the number of components required. An example of this is shown in Figure 32. This circuit resists reduction through further manipulation, yet its function can be duplicated by the Not element alone when the second input is used as a variable (inhibit function; see Figure 33). This circuit is also called a nonimplication circuit and is a common circuit combination. Others will be shown in the next section.

The Not element using both inputs is shown in the circuit diagram in Figure 34. In this application, the cylinder will extend only if push button B is actuated and input A is *not* actuated.

Design Notes: The port designations of the Not element must be clearly marked, when both inputs are used, for the function to be understood. The inputs may approach the Not element from any direction (see Figure 35). It is also proper to substitute an I in place of an N on the Not element to show an inhibitor function (Figure 36). In actual practice this is, however, seldom done. Some designers contend that this can be confus-

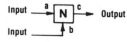

Figure 30

a	b	c
0	0	0
1	0	0
1	1	0
0	1	1

Figure 31

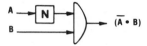

Figure 32

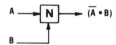

Figure 33

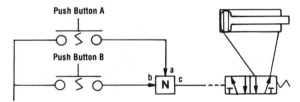

Figure 34

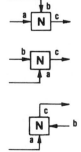

Figure 35

42 True Logic Elements Or-And-Not

Figure 36

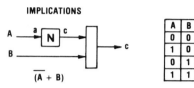

Figure 37

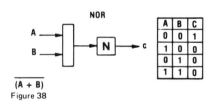

Figure 38

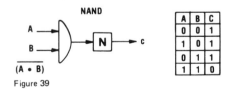

Figure 39

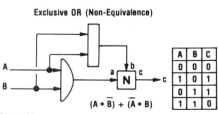

Figure 40

True Logic Elements   43

ing because it makes the drawing appear to contain two different types of hardware when actually the Not element is used for both functions.

Common Combinations Using the OR, AND, and NOT

The following illustrations show common combinations of the elements Or, And, and Not. These combinations are common enough that they are given their own names. Remember that the Not element when shown with only one input is used as an inverter.

The implication circuit shown in Figure 37 is also sometimes called an Or-Not circuit. Notice that the output conditions are quite different from that of the nonimplication (And-Not) function previously shown in Figure 33.

The Nor function shown in Figure 38 and the Nand function shown in Figure 39 are simply the Or and the And function with their outputs inverted.

The exclusive Or or nonequivalence circuit shown in Figure 40 functions exactly like an Or element except that in the final condition, where both inputs are on, the output is now off. Nonequivalence means that the output is on whenever the two inputs are not equal to one another.

The opposite of the exclusive Or is shown in Figure 41, the equivalence circuit. Notice here that the output of this circuit is on only when both inputs are equal (both on or both off).

All of these circuits are shown in their simplest form. Each can be expanded to include a greater number of inputs.

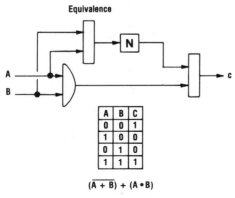

Figure 41

5
TIMING FUNCTIONS

BASIC PRINCIPLES

Pneumatic timing functions are created in a variety of ways. The most common type and the type used to describe timing functions in the ANSI Standard is the use of an orifice and an accumulator to delay the response of a snap-acting element. Snap acting means that the element will shift quickly and consistently when a given pressure is applied to the input. The diagram in Figure 1 shows the orifice and accumulator connected to an And element.

The flow diagrams in Figure 2 show the action of this circuit. The orifice/accumulator causes a slow build-up to the shift point of the snap-acting element. The time delay is created between the time that the signal is applied to the input and the time that the output goes on.

Most timers are adjustable. This can be accomplished by adjusting the size of the orifice or by changing the size of the accumulator. The arrow in the diagram in Figure 3 indicates an adjustable orifice is used.

You will also note the addition of a check valve to the drawing in Figure 3. The function of the check valve is to allow the air to bypass the orifice when the input signal is removed. The timing assembly allows you to create a time delay when the signal is applied, but with immediate response when the signal is removed, as shown in the graph in Figure 4.

Basic Principles

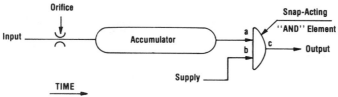

Figure 1

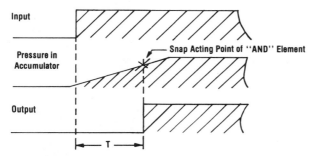

Figure 2

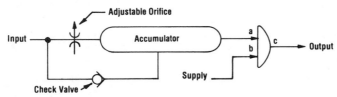

Figure 3

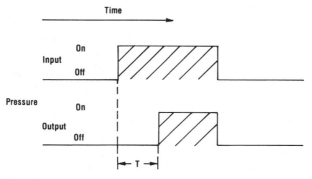

Figure 4

46   Timing Functions

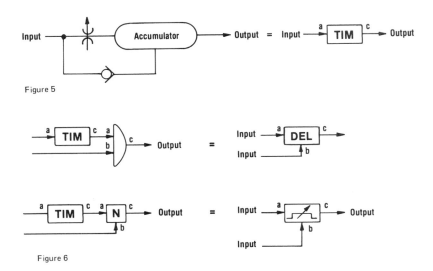

Figure 5

Figure 6

TIMER, DELAY, AND PULSE ELEMENTS

The adjustable orifice, accumulator, and check valve assembly can also be shown using its logic symbol called a "timer" element (Figure 5). This timer element must always be connected to a snap-acting element.

The timer element is often assembled to the snap-acting element it will actuate. When these two components are assembled into one element, they are called a "delay" (if mounted to an And element), and a "pulse" (if mounted to a Not element). Figure 6 shows these relationships.

BASIC TIMING FUNCTIONS

The delay and the pulse element can be used to create the eight basic timing functions shown in Figures 7 through 14. Timing in is the most common of all timing functions. All timing in functions indicate that the timing takes place when the input signal is applied (Figure 7). Timing in inverted is the same timing function as timing in except that the output is normally on (for inverted, see Figure 8).

Timing out means that the timing is done after the input signal is removed. An additional Not element is used to invert the signal to the pulse element (Figure 9). In Figure 10, timing out inverted again is the same as timing out except that the output signal is normally on.

Timing in and out creates a time delay both at the time when the input goes on and when it is removed. These times can be independently adjusted (Figure 11).

Basic Timing Functions 47

The pulse function converts a signal of longer length (in time) to an output of a specific duration. Notice that the a and b ports of the pulse are both connected to the input (Figure 13). The pulse-inverted function is again the reverse output condition of the pulse function (Figure 14).

These are the eight standard timing functions. Many additional examples of applications involving timing are given in Chapter 6.

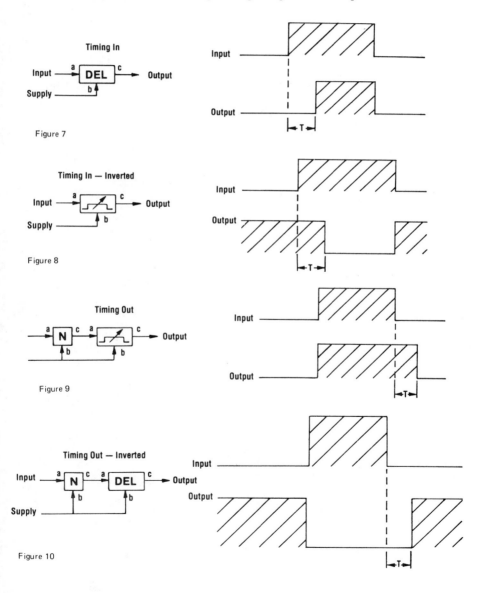

Figure 7

Figure 8

Figure 9

Figure 10

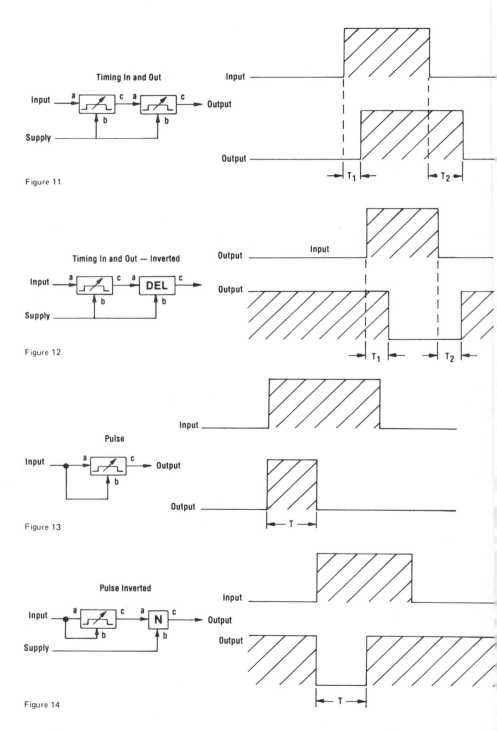

Figure 11

Figure 12

Figure 13

Figure 14

6
MEMORY FUNCTIONS

BASIC PRINCIPLES

Memory functions are the heart of most control circuits. Only memory functions have the ability to "store" information. This storage capability is used in all circuits that are sequential in nature. The memory records the functions of the machine or process as they occur. Then they direct new activities and provide proper interlocks, all based on the information they have received and the design of the circuit itself.

The basic function of a memory, whether it is called a memory, a flip flop, or is a "memory circuit" made up of standard elements, is always the same. Using the standard memory symbol, the basic function of a memory is shown in Figure 1.

With *supply on*, the memory functions as follows:

1. When the set input goes on (pressurized), the memory output shifts on (pressure).
2. The memory will remain in this condition (memory set, output on) even though the set input is removed.
3. When the reset input goes on (pressure), the memory output shifts off (no pressure).
4. The memory will then remain reset (output off) even though the reset input is removed.
5. Until a set input is again received.

Thus the memory "remembers" the last signal it was given, either to set (output on) or to reset (output off). Provided a constant supply is main-

50   Memory Functions

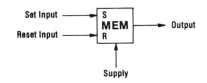

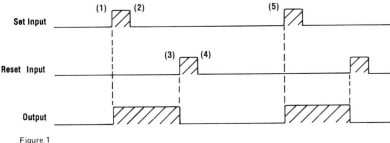

Figure 1

tained, there is no known limit to the length of time the memory can "remember" the last command it was given.

These functions are common to all memory devices. Were it not for other important characteristics, one memory could be used for all applications. There are, however, other important considerations involved in selecting the type of memory that will best suit each application.

VARIATIONS IN MEMORY FUNCTIONS

Figure 2 illustrates a memory function with a set and a reset output. This memory circuit provides the designer with two output signals: one which reflects the set input and a second which reflects the reset input.

Some types of memory functions depend on the air supply to hold their set or reset condition—others do not. A typical memory which depends on supply will shift to the reset mode when the supply is removed and reapplied. This is shown in Figure 3.

The third factor involves the reaction of the memory to both inputs present at the same time. Figure 4 illustrates this condition. Several possibilities exist for this condition. They are: (1) both outputs on, (2) both outputs off, (3) go to set mode, (4) to go reset mode, and (5) first input to arrive has control.

The fourth consideration is the response time. Shift points and flow capacity will vary depending on the memory selected. Often this requires a

Variations in Memory Functions

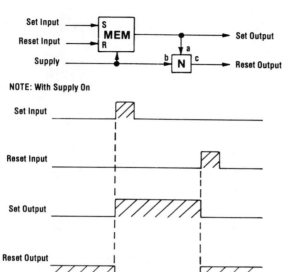

Figure 2

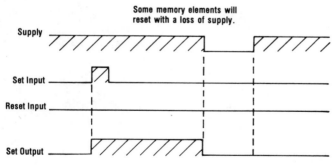

Figure 3

52 Memory Functions

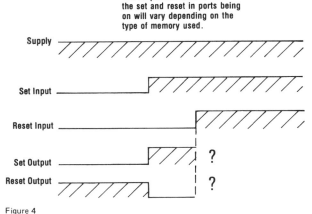

Figure 4

close examination of the components used. The following 10 examples will illustrate the wide variety of pneumatic memory circuits available and list some of the characteristics that may cause them to be selected.

MEMORY CIRCUITS USING TRUE LOGIC ELEMENTS

The memory circuit shown in Figure 5 functions as follows:

1. Supply is present at the b port (the supply port) of the And element.
2. A momentary signal at the set input causes the output of all three elements to go on (memory will set).
3. Once set, this memory will hold set due to the connection from the output to the b port of the Or element.
4. A momentary signal at the reset port will cause this memory to reset. The signal at the a port of the Not element causes all element outputs to go off (exhaust).

The individual characteristics of this memory circuit are:

1. Number of outputs—set output only.
2. Effect of loss of supply—memory will return to reset condition.
3. Effect of set and reset signals present—memory will be reset, output off.
4. Other factors—response times—should be good. Snap action is dependent on And and Not elements used.

Memory Circuits Using True Logic Elements 53

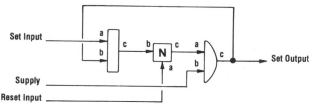

Figure 5

Other options:

1. A reset output can be added by connecting a second Not element to this memory circuit, as shown in Figure 3.
2. Memory circuits with "normally on" reset signals can be constructed by making modifications to this circuit. The diagram and charts in Figure 6 show this variation. This circuit is accomplished by combining the reset and supply inputs. The alternate circuit shown in Figure 7 will provide a much quicker (snap-acting) reset function.

Figure 8 shows a second basic type of memory circuit using standard elements. The characteristics in this circuit are:

1. Outputs—both set and reset outputs provided.
2. Effect of loss of supply—position of memory cannot be predicted upon return of supply (see circuit option 1).
3. Effect of set and reset signals present—*no outputs* (see option 2).

Option 1: This circuit can be made to assume the set or reset position after a loss of supply by adding a delay to the b input of the set or the reset Not element (Figure 9). The memory in Figure 9 is now designed to return to the reset condition when supply is removed and reapplied.

Option 2: This circuit can be designed so that any of four different output conditions can be achieved when both the set and reset signals are present.

The circuit in Figure 10 will have both outputs *on* when the set and reset are both present. The circuit in Figure 11 will have the set output *off* and the reset output *on* when the set and reset inputs are both present. The circuit in Figure 12 will have the set output *on* and the reset output *off* when the set and reset inputs are both present.

Thus, with this circuit, you can actually design the relationship you wish to occur in the event that both inputs are on at the same time.

54 Memory Functions

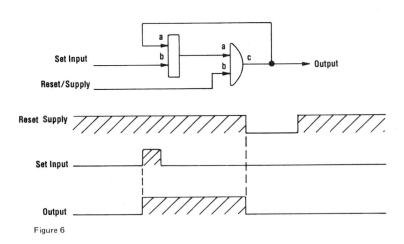

Figure 6

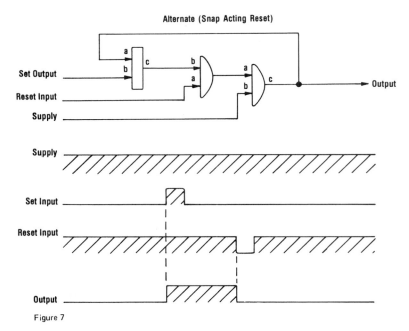

Figure 7

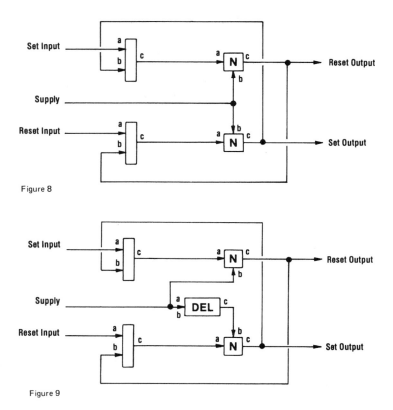

Figure 8

Figure 9

SPECIAL MEMORY ELEMENTS

Exploring all of the possible special valve arrangements which can be used to create memory functions would be practically impossible. However, the two examples shown in ANSI Standard will serve well to demonstrate the possibilities. Figure 13 illustrates the memory element shown in the ANSI Standard.

Figure 14 shows a more detailed breakdown of the pilot-operated check valve which may be helpful to you in following the operation of this memory. The description of this function is as follows:
1. With supply on:
2. Set signal passes through check valve and shifts three-way valve. Set output now on.
3. When set signal is removed, pressure is trapped between the check valve and the pilot port. Valve remains shifted (set). Ori-

56 Memory Functions

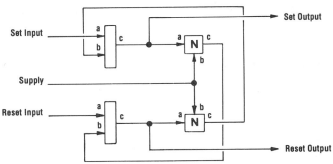

Figure 10

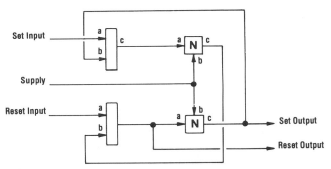

Figure 11

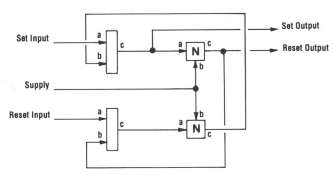

Figure 12

Special Memory Elements   57

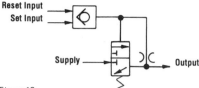

Figure 13

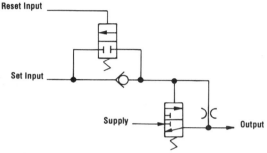

Figure 14

fice connecting output to pilot maintains the pressure at the pilot. This orifice prevents small leaks from affecting the memory function.
4. The reset input shifts the two-way valve to a passing condition. This allows the pressure trapped between the check valve and the pilot of the three-way valve to exhaust to atmosphere (faster than it can be made up by the orifice). This causes the three-way valve to return, exhausts output, and the memory is now reset.

Characteristics of this memory are:

1. Outputs—set only; requires an additional component to provide reset output.
2. Effect of loss of supply—memory will return to reset condition.
3. Effect of set and reset signals present—memory will be set, output on.
4. Other factors—response should be reasonably good. Set will not be snap acting (due to orifice). Reset can be snap acting.

Flip Flop Element

The flip flop device shown in the ANSI Standard is an example of a memory function which is mechanically retained. The symbol for the flip flop and the valve drawing for this device are shown in Figure 15. The valve

58 Memory Functions

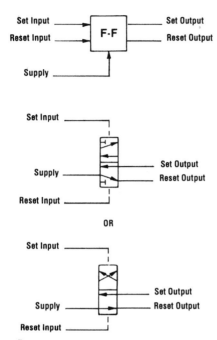

Figure 15

symbol (either the five-or four-ported four-way valve) shown in Figure 15 is used to illustrate the function of the flip flop device.

1. With supply on, the set output or the reset output will be off.
2. If the reset output is on (and set off), a signal at the set input port will cause the outputs to reverse (set output on, reset output off).
3. The flip flop will remain in this condition until (1) the set input signal is removed, and (2) a signal at the reset input is received.

Here are the individual characteristics for a flip flop function:

1. Two outputs:
2. Effect of loss of supply—the flip flop function is mechanically retained. Its storage capability is not affected by power interruptions. Thus, when supply is returned, the flip flop will reflect the position it was last commanded to assume (Figure 16).
3. The effect of set and reset signals present—the flip flop typically satisfies the fifth and final possible reaction to this condition in

Special Memory Elements 59

that the first signal to arrive at the flip flop (either set or reset input) take control. For example, if the set signal is applied first and later the reset, the flip flop will remain in the set mode. The reverse is also true. If the reset signal is applied first and the set second, the flip flop will remain in the reset position. The graph in Figure 17 illustrates this function.

In summary, the following characteristics of the flip flop should be remembered:

1. Loss of supply does cause the output of the flip flop to go off. However, loss of supply by itself does not change the condition of the flip flop.
2. The condition of the flip flop can be changed by a set or reset signal with supply off.
3. If a set signal is applied first and held, a reset signal will not change the output condition of the flip flop.
4. If a reset signal is applied first and held, a set signal will not change the output condition of the flip flop.
5. Statements 3 and 4 are with supply on or off.
6. Statements 3 and 4 assume that the set and reset signals are of equal pressure.

Although all types of memory devices are useful in certain applications, the flip flop with the characteristics just described is the memory device most often selected for sequential circuit design.

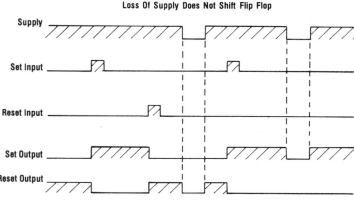

Figure 16

60   Memory Functions

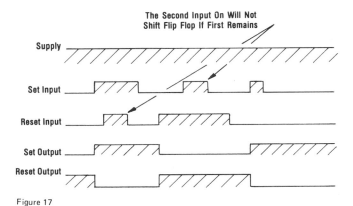

Figure 17

COMMON CIRCUITS USING MEMORY AND DELAY FUNCTIONS

Like the common circuits which were shown for the Or, And, and Not functions, there are also common circuits which use the functions of delay and memory. These circuits are either so common or so unique that they often have names associated with the functions they perform.

Signal Standardizer Circuit

The function of the signal standardizer circuit is to provide a fixed signal duration disregarding the duration of the input (Figure 18). You will notice that the first input is shorter and the second is longer than the desired output time. Nevertheless, the circuit will respond by producing the output adjusted at the delay element. To accomplish this function, you must use a memory which has set priority or a Flip Flop memory function. This is because the set output of the memory must no-go off before the input is released.

This circuit is often used to correct for possible operator error, such as holding a button or foot pedal down too long or too short a time. Other applications occur when a limit valve will be actuated for different periods of time, depending on machine speeds.

Two-Hand Anti-Tie-Down Circuits

To understand the function of the two-hand anti-tie-down in Figure 19, it is helpful to study its individual parts. The two Not elements shown in Figure 20 form a priority relationship in that if input 1 arrives first, it will prevent output 2 from going on. The reverse is also true—if input 2 arrives first, it will hold output 1 off.

Common Circuits Using Memory and Delay Functions 61

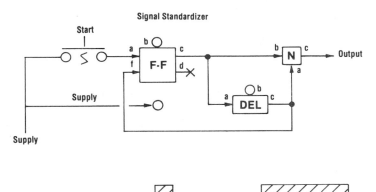

Figure 18

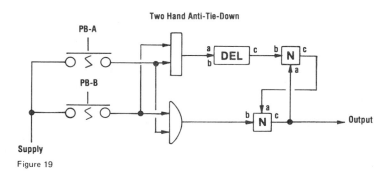

Figure 19

The second relationship is that of the Or and the And elements. Notice in Figure 21 that the Or element output goes on when either input goes on. The And element output goes on only when both inputs are on.

The delay is used to complete the circuit. The delay is started when either A or B go on. If both A and B are present before the delay is complete, the output of the circuit will go on. If both A and B are not present before the delay is complete, the output will not go on. This function is shown in Figure 22.

The most important function of the two-hand anti-tie-down, how-

62   Memory Functions

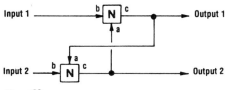

Figure 20

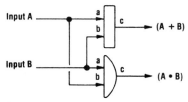

Figure 21

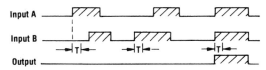

Figure 22

ever, is what happens when one hand is released. The graph in Figure 23 illustrates this function. Notice that if either input is released, the output will go off and remain off until both inputs are released and again initiated together. This is the anti-tie-down portion of the circuit. It means that both inputs must be off between cycles. This is insurance against not only intentional acts, such as tying down one of the buttons, but also and probably more important, an undetected malfunction of the push-button valve itself.

Oscillator Circuits

Oscillator circuits are another very common circuit requirement. Figure 24 shows one example of an oscillator circuit. This oscillator will perform as follows. When the input (which also serves as supply in this circuit) goes on, output 1 goes on. Output 1 remains on for the period of time adjusted at delay 1 (T1). After the delay, output 1 goes off and output 2 goes on; output 2 remains on for the period of time adjusted at delay 2 (T2). When this delay is complete, output 2 goes off and output 1 goes back on. This sequence will repeat until the input/supply is removed.

Common Circuits Using Memory and Delay Functions

Another example of an oscillator circuit is shown in Figure 25. This oscillator circuit will operate the same as the oscillator shown in Figure 24 except that when the input/supply goes on, either output may be first the on. This will depend on which output was on when the input/supply was removed.

There are many requirements for oscillator circuits, and in addition, more elaborate oscillator circuits involving more delays and more outputs are also common. These will be covered more thoroughly in Chapter 8.

Alternate Timing Out

The circuit shown in Figure 26 can be used as a substitute to the timing out function shown in the timing section. Unlike the timing out function shown in the timing section, this circuit will not create a pulse when supply goes on.

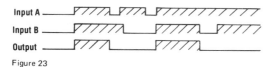

Figure 23

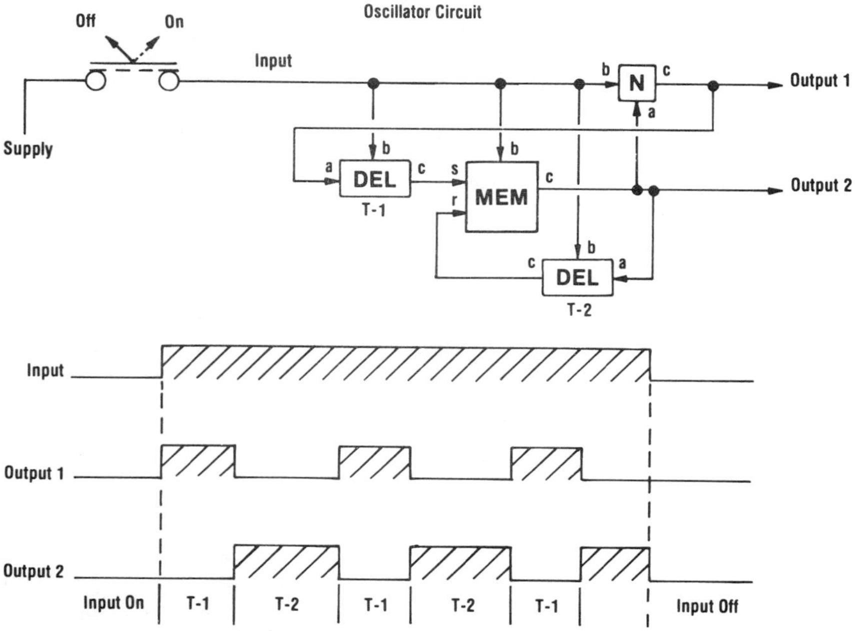

Figure 24

64 Memory Functions

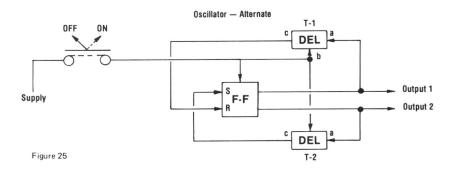

Figure 25

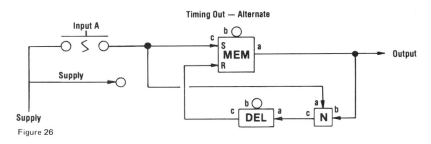

Figure 26

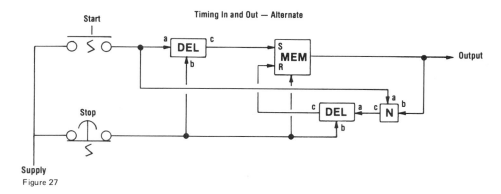

Figure 27

A timing circuit similar to this might be used for the vent fan in a paint booth. If the delay were set for 5 minutes and the input was pressurized when the paint gun was operated, the fan would turn off 5 minutes after the gun was last operated.

Additional variations of this circuit, such as timing in and out, can also be designed. Figure 27 illustrates just one of the timing circuits possi-

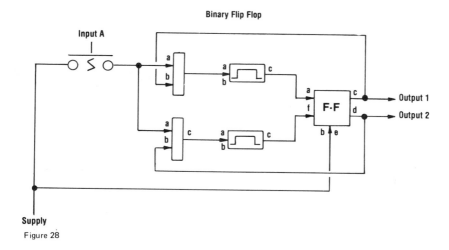

Figure 28

ble using variations of the timing out circuit. This one is timing in and out with an emergency stop option.

Binary Flip Flop Circuit

Figures 28 and 29 show binary flip flop circuits and describe their functions. These binary flip flops function because of the pulse elements which deliver only one pulse when actuated and held actuated. Once the flip flop is in either the set or reset position, it holds the pulse element in that position pressurized. Thus, when a new input is sent to both Or elements, only one pulse element will produce an output signal. Each time an input is received at A, the flip flop will shift one position: if it is set it will reset; if it is reset, it will set.

The applications for this type of circuit are not uncommon. It might be used as a sorter circuit to equally distribute material from one conveyor to two conveyors. In multiples, they can also be used for counting.

A composite symbol is sometimes used to show the binary flip flop and other complex functions. This symbol is a rectangle with two additional vertical lines, as illustrated in Figure 30. The symbol is then identified with letters or numbers which help to indicate its function.

As of this writing, this symbol has been formally proposed as an addition to the Standards; however, the letter or number codings have not been established for each device. When using these symbols, be careful to include a description of the function of the device as a reference on your drawing.

66 Memory Functions

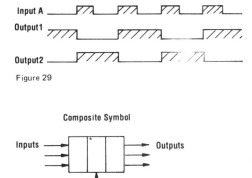
Figure 29

Figure 30

Figure 31 shows a drawing of binary flip flops which form a counting circuit using composite symbols.

R-S-T Flip Flop

An R-S-T flip flop is a device which combines the function of a standard and a binary flip flop. It has a supply and three inputs. Two of these are the set and the reset inputs which cause the element to shift exactly as would the inputs to a standard flip flop. The third input (called a trigger) causes the element to shift one position (binary flip flop function). Figure 32 shows a functional diagram and a composite diagram of an R-S-T flip flop.

Shift Register Circuit

The shift register circuit is used to store and transfer information in the order it was received (see Figure 33). An example of this would be a car wash. When the car enters the car wash, the operator pushes the button for either a hot or cold wax. This information is not immediately used since there may be several cars between the entry area and the wax station, each with its own requirements as to hot wax or cold wax. Therefore, this information must follow the car and be given to the wax unit just as the car enters the wax area.

Similar situations arise in machine control circuits. For example, a part may be tested in one station and not rejected or accepted until several stations later. A shift register will perform this function. A stage of a shift register consists of two flip flop elements. It is important that inputs C and

Common Circuits Using Memory and Delay Functions 67

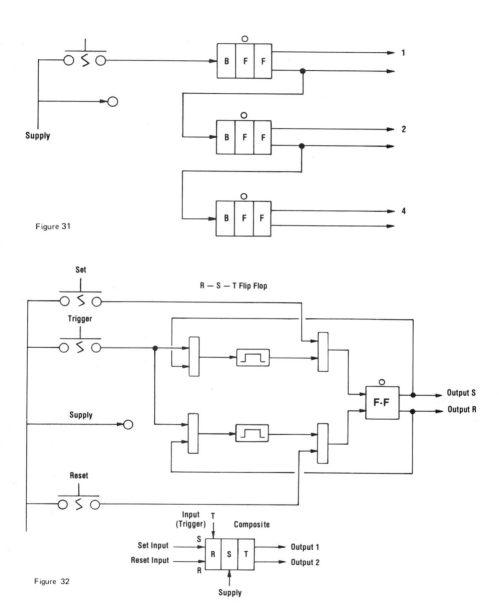

Figure 31

Figure 32

68   Memory Functions

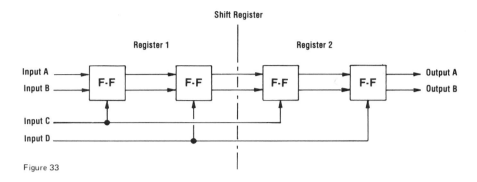

Figure 33

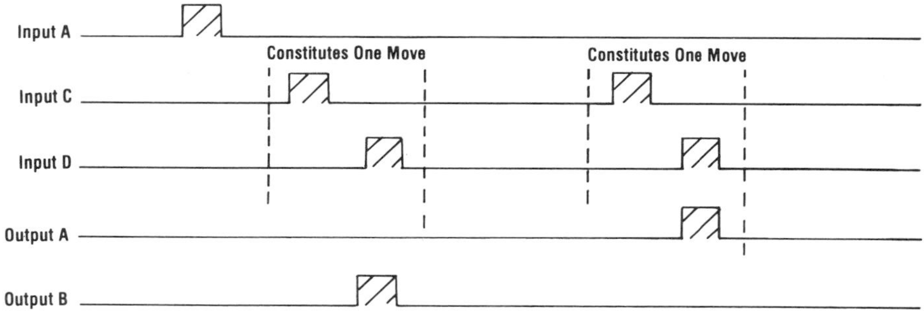

Figure 34

D are not on at the same time. Assuming that all of the flip flops are shifted to the reset position, the graph in Figure 34 will illustrate the action of getting a set signal to the final position.

Figure 35 shows additional composite symbols and identifies them by the functions they perform. Composite symbols can be used for two purposes. One is to make the drawing more understandable by eliminating much of its complexity. This "concept" drawing can often be useful to clearly illustrate the function of the overall system. The second use for these symbols is in cases where the manufacturer has chosen to combine the individual functions necessary to create the complex function into one body. In this case, the concept drawing can be the working drawing as well.

Common Circuits Using Memory and Delay Functions 69

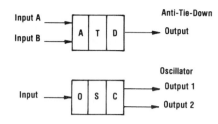

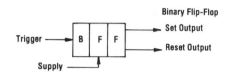

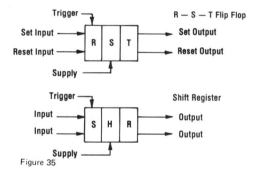

Figure 35

7
SPECIAL-PURPOSE ELEMENTS

BASIC PRINCIPLES

The fourth group of control components are the special-purpose components. Most of these components, unlike the others previously described, change the nature of a signal and are called interface elements or components. They convert an electrical signal to an air signal, low-pressure air to high-pressure air, vacuum signals to pressure signals, and so on. These devices allow the pneumatic control to "talk" to other types of control systems and allow the other systems to "talk" back. Some of these devices are very common, such as an orifice or a regulator. Others are more unique to air control circuits and will require additional explanation.

ELECTRICAL TO AIR INTERFACE

Electrical to air interface is accomplished by a solenoid-operated air valve, generally operated directly by an electromagnetic coil and returned by a spring. The basic symbol for a solenoid coil is shown in Figure 1. This symbol can be used with the terminal and bridge to show solenoid valves as inputs to a control circuit. The symbol can also be combined with standard logic symbols to show a solenoid valve as a part of the control circuit.

AIR TO ELECTRICAL INTERFACE

Air to electrical interface is accomplished by a pressure-operated electrical switch. This is a set of contacts (in some cases several sets) operated by an

Air to Electrical Interface

air piston or diaphragm. Some pressure switches have adjustable shift points both on and off. These pressure switches can be used as an air/electrical interface; however, they are much more sophisticated than is actually required. The pressure switch here is used merely as a converter from an air to an electrical signal. Symbols used for air to electrical interface are shown in Figure 2.

The symbol at the top of Figure 2 can be used to represent a pressure signal in an air circuit diagram. The symbol at the bottom of Figure 2 shows more details of the switch function as they would appear in the electrical diagram.

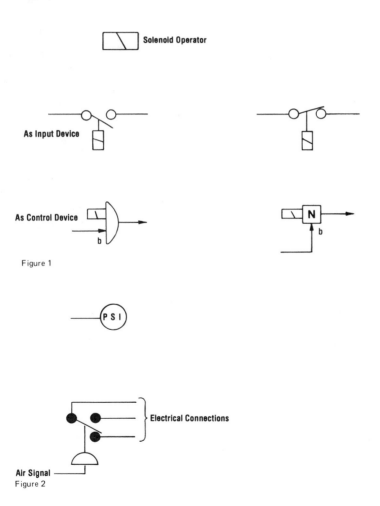

Figure 1

Figure 2

HIGH TO LOW PRESSURE

A simple air regulator is used to reduce the output pressure from the pneumatic controls (see Figure 3). The pressure interface high to low would be used to allow pneumatic controls to communicate with fluidic functions (nonmoving part logic) and other types of controls which operate at lower pressure.

LOW TO HIGH PRESSURE

Low to high pressure interface is accomplished with an amplifier element. An amplifier element could be any valve which can control a higher pressure than is required at the pilot port. For example, if the valve in Figure 4 can control a 100 psi signal with a 25 psi pilot signal, this valve would also become an amplifier with a value of 4 to 1. Thus, practically all valves offer some degree of amplification.

Amplifiers in air circuits are commonly used for two purposes. The first is to convert low-level signals from fluidic devices or other low-pressure controls. Since these controls produce a signal of approximately 3 to 5 psig, an amplifier with a valve of approximately 50 to 1 would be sufficient to convert these signals to moving-part logic control pressures. Symbols for these amplifier devices are shown in Figure 5.

The second application for amplifiers in a pneumatic circuit is to amplify signals that are returned from noncontact sensors. These sensors were described in Chapter 2. The signals produced from these sensors are often as low as 3" H_2O and require much more amplification than the fluidic signal (1" H_2O equals approximately .038 psi).

To amplify a signal of .1 psi to 100 psi, an amplifier of at least 1,000

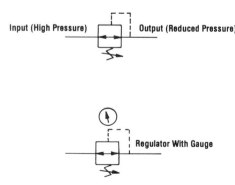

Figure 3

Flow Reduction and Increase 73

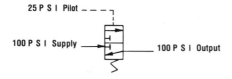

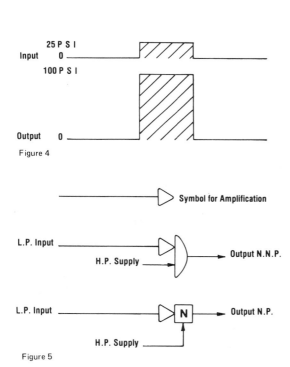

Figure 5

to 1 amplification would be required. In cases such as this, you will often find that the signal from the sensor is amplified twice, as shown in Figure 6. Thus two amplifiers, both with 50 to 1 ratios, designed to operate with different supply pressures, can do the work of one extremely sensitive amplifier when they are connected in series as shown.

FLOW REDUCTION AND INCREASE

The function of flow reduction is performed by an orifice. The diagrams for adjustable and nonadjustable orifices are shown in Figure 7.

The function of increased flow is performed by a pilot-operated

74 Special-Purpose Elements

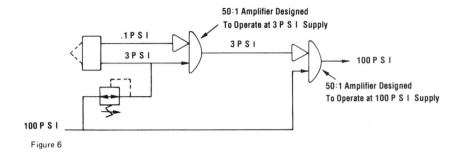

Figure 6

valve. This is simply a larger-capacity valve coupled to a large-capacity supply. This is the function of most output valves (called power valves) in the control circuit (Figure 8).

The power valves on most air control circuits actually can perform several functions at the same time. The four-way power valve shown in Figure 9 is performing the following functions:

1. Converts a three-way to a four-way signal.
2. Amplifies the signal 2 to 1.
3. Amplifies the flow 5 to 1
4. Isolates lubricated air from the control circuit.

THREE-WAY TO TWO-WAY INTERFACE

Conversion of the three-way signals produced by the air controls into a two-way signal can be accomplished in several ways. Figure 10 illustrates some of these. Two-way valves can also be used to convert three-way signals to two-way or bleed signals.

TWO-WAY TO THREE-WAY INTERFACE

As described in Chapter 2, bleed valves (two-way) are sometimes used because they are smaller, have only one tube connection, or are easier to operate than three-way valves. The two-way signal produced by these valves must then be converted to a three-way signal for use in the air logic control system. The symbol and basic function of the two- to three-way converter are shown in Figure 11. Supply pressure is fed through an orifice to the a port of the Not element and the bleed valve. With the bleed valve closed, pressure builds to supply pressure on the a port of the Not element (output off).

Two-Way to Three-Way Interface 75

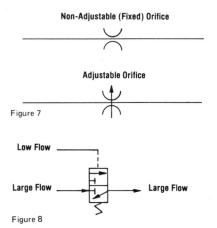

Figure 7

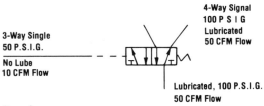

Figure 8

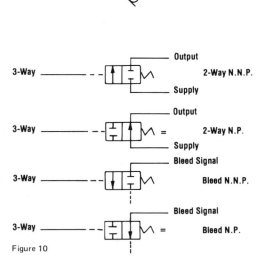

Figure 9

Figure 10

76   Special-Purpose Elements

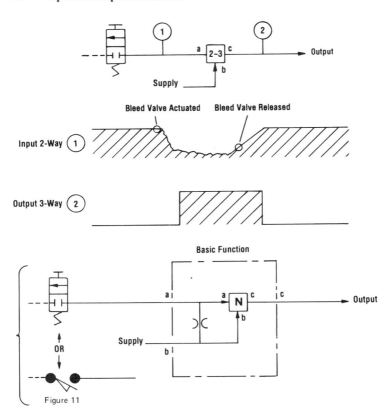

Figure 11

When the bleed valve is actuated, the air pressure is exhausted to atmosphere faster than it can be made up by the orifice; pressure drops at the a port of the Not element, and the output of the Not element goes on. When the bleed valve is released, pressure again builds at the a port of the Not element and the output of the Not element again goes off.

When using bleed valves, you should be conscious of the limiting factors inherent to their use. They are:

1. The response time will be greater in most cases than those of a standard three-way limit valve.
2. Additional length of tubing between the bleed valve and the converter will accentuate this problem.
3. A broken connecting line between the bleed valve and the converter will cause the converter output to go on.
4. The output of the converter will pulse when the air supply is turned on to the circuit.

AIR TO HYDRAULIC

Pneumatic signals can be used to control hydraulic cylinders and motors. The interface is performed by an air pilot operated hydraulic valve. A typical application is shown in Figure 12.

HYDRAULIC TO AIR

Hydraulic to pneumatic interface is accomplished by the use of pressure, level, or flow sensing valves. The diagrams for these devices are shown in Figure 13.

In many cases, only certain types of these devices are available in hydraulic to air models. In cases where they are not available, it may be necessary to use an electrical pressure switch or flow switch and later convert these signals to pneumatic signals through the use of a solenoid valve.

AIR TO VACUUM

Pneumatic signals can be converted to vacuum signals by either opening a vacuum valve or by supplying a venturi type of vacuum generator (Figure 14).

VACUUM TO AIR

Vacuum signals can be converted to pneumatic signals by the use of a vacuum sequence valve. A vacuum sequence valve can be one that is designed specifically for this function or, in some cases, it is simply an overadjusted amplifier element. The symbol for a vacuum sequence valve is shown in Figure 15.

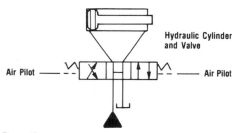

Figure 12

78   Special-Purpose Elements

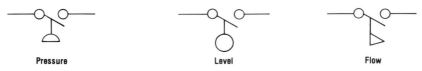

Pressure   Level   Flow

Figure 13

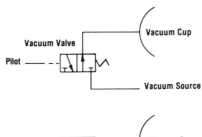

Figure 14

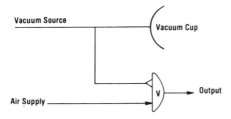

Figure 15

8
SEQUENTIAL CIRCUIT DESIGN BASIC PRINCIPLES

INTRODUCTION

In the previous chapters we have used many illustrations of circuits. Many of these were combinational logic circuits. Combinational logic circuits respond to a combination of input signals. When the proper combination exists, the output will be on. When this combination does not exist, the output will be off. The order in which these input signals occur is of no consequence.

At the heart of most machine and process controls, however, is a sequential circuit. These machines or processes require that their actions take place in a specific sequence or pattern. For safety and efficiency, these controls must be designed to accept and react to signals only if they arrive in the proper sequence. Figure 1 graphically depicts and lists some of the differences between the two basic types of circuits.

Because the sequential circuit is the heart of most controls, the circuit designer would be lost without a good method for designing these controls. In the following chapters, one method is described in detail, with many illustrations on various circuits. This method was selected because of its ability to fulfill the requirements of all five good circuit design practices. These practices should be studied carefully because: (1) they are used to judge alternate circuit design methods, and (2) every circuit you design should be examined carefully to be sure you have fulfilled the requirements in each of the five categories.

80 Sequential Circuit Design

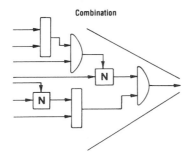
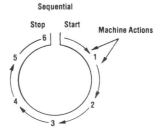

Combination

A — Responds to a combination of inputs received in any order
B — Outputs produced in any order
C — Has no real beginning or end
D — Uses only Or, And, Not
E — Can be described using algebra expressions and factored

Sequential

A — Responds only to inputs received in the proper sequence
B — Outputs must be produced in a specific sequence
C — Has definite beginning and end
D — Uses all elements
E — Designed by using sequential logic methods

Figure 1

Good Circuit Design Practices

Many circuit designs fail because the designer does not have a thorough understanding of good circuit design practices. Thus, the designer will select a method and, ultimately, designs which do not fulfill all of the requirements of good circuit design. The requirements are as follows:

1. Function
2. Safety
3. Reliability
4. Repairability
5. Cost

These five requirements should be uppermost in the mind of the designer when selecting the method to be used, when designing the circuit, and when checking the circuit before use. Unfortunately, all too often designers recognize only some of these requirements (usually 1 and 5). They design and check the circuits for the requirements they understand, and the rest is left to chance. This can lead to some disastrous results. Here is a description of each requirement.

1. *Function* is generally well understood. You must achieve the desired sequence of operation or the circuit will be of no value. Obviously, the method you use must deliver a working circuit functionally over a wide

variety of applications. It must also be simple enough that it can be easily understood and used.

2. When designing a control circuit, *safety* is of the upper most importance. Some safety requirements are very apparent, others are not. For example, machine start interlocks and emergency stop requirements are often apparent and given much consideration when the circuit is designed and checked. On the other hand, the actions that could occur due to an accidental actuation of a limit valve or push button, a malfunction of these devices, a loss of control pressure, and other such occurrences are not very apparent and are often not considered by the designer.

The method used to design the sequential circuit can be helpful here. The method should deliver the desired sequence of operation *plus* prevent any sequence other than the desired sequence from occurring. Thus, the method itself can become an aid to safety by automatically including interlocks which might otherwise be overlooked. In addition, it is important that the designer's role in machine safety be understood and appreciated. Only then can we expect that the designer will take proper care to examine the results of each design for safety.

In examples that will follow later, we will point out some of the less apparent safety problems and how they can be controlled.

3. *Reliability* is one of the circuit designer's contributions to the productivity of the machine. Many designers believe that once they have selected the most reliable hardware to use, their contribution to reliability is completed. True, reliable hardware is important, but what they do not realize is that the most reliable hardware does not guarantee a reliable circuit. Much of what makes a circuit reliable is the design of the circuit itself. So, the method used to design this circuit should do two things. First, it should be oriented toward basic functions. This allows the widest variety of product selection. In other words, the design method should not be geared around a special component that only one manufacturer can provide—this may not be the brand you would select for every application. Second, the design method should be geared toward a circuit design which will operate properly over the widest range of conditions possible. Marginal elements and combinations of elements must be avoided. For example, pulse elements will not be called out by a design method which considers reliability. Pulse signals severely limit the conditions under which a pneumatic circuit can operate, often reducing the overall dependability of the circuit. These devices should be included by the designer only after careful consideration of their effect. A skilled designer will also review the overall designs, specifically looking for weaknesses in dependability and ways to improve the dependability of every circuit.

4. *Repairability* is the second contribution made by the designer to the productivity of the machine. Here the designer must take a different view of the circuit and assume that, in spite of all efforts to design a dependable circuit, at some point during the life of the machine every component will fail. With this in mind, the designer must review the circuit two ways. What can be done to help identify the problem? What can be done to help repair personnel correct the problem? The circuit itself can be designed to aid in identifying the problem. Many times simple diagnostic tools such as a gauge or an indicator can be of valuable assistance if they are considered and included by the designer. The design method used is also important to repairability. For example, pulse signals eliminated for dependability reasons are also very difficult to troubleshoot. Maintained signals are much more valuable in determining the cause of a problem.

Review by the designer with an eye toward troubleshooting and repair are tremendously important. Included in this is the proper documentation of the circuit and its components, as well as a recommended spare parts list.

5. Let us not forget *cost*. The designer's responsibility is to deliver a circuit which incorporates the proper amounts of function, safety, reliability, and repairability at the lowest possible cost. Again, the design method and the designer's skills work in concert to accomplish this task. First the design method, by calling out basic functions, can allow the designer the widest selection of components. If it is a well-planned approach, it should also be helpful in reducing unnecessary redundancies. The designer's skills are used in reviewing the circuit, carefully selecting the proper hardware to be used, and weighing the cost of each component against the values received in function, safety, dependability, or repairability. In the final analysis, the designer should be able to explain the function and the alternatives of each component as well as the gains or losses to each of the categories listed as a result of its use.

This sounds like a large order but in fact, due to the similarities of circumstances in many designs, much of this will, through experience, become "second nature" to the designer. Once the designer has studied an application thoroughly, many conclusions will carry over into the second design, the second to the third, and so on. Soon a pattern will have developed and the designer need only look closely at the difference between this and previous experiences. For this reason, it is important that the designer first completely understand the requirements of good circuit design and, second, learn a sound design method thoroughly.

The first example in the next section explains the design method and reviews each of these requirements.

SEQUENTIAL CIRCUIT DESIGN METHOD

The circle at the top of the sequential description in Figure 1 is representative of a sequential machine sequence. Most machine cycles can be represented as a circle, with the start and stop points adjacent points on the circle. Once started, the automated machine follows around the circle, fulfilling the requirements of its cycle until it arrives back where it started, the "end of cycle." The circle is made up of segments. These segments are one action the machine must do. These segments drive one another; in other words, segment 1 must be completed before segment 2 can start. Segments 1 and 2 must be completed before segment 3 can start, and so forth. Thus, each segment is based on the preceding "history" of the cycle.

The word "history" is important in this type of circuit because it indicates that the circuit must record the actions of the machine as they take place and create this history as it directs new actions. This is precisely what happens in a well-designed sequential circuit.

The actions of the machine are signaled to the control circuit by the use of input signals beginning with the signal which starts the machine cycle. The information which will be lost before the end of the cycle (such as from limit valves that will be released before the end of the cycle) is recorded by using memory functions. Information which will not be lost, such as limit valves which will be maintained actuated until the end of the cycle, will be combined with the outputs of the memory function by using an And element. Thus, a complete record is kept of the sequence of operations within the circuit. The record is maintained until the end of cycle occurs. At this point, all of the history of the cycle is erased in preparation for the new cycle to occur. This is done by resetting all set memory functions.

To illustrate the method used to design a sequential circuit, we will use the following example.

Sample Problem Design

This is a circuit for a machine which punches a hole in a part. It is loaded and unloaded by the operator. Only two air cylinders are involved in the actions of the machine—one which holds the part in place (a clamp cylinder) and one which operates the punch. The air cylinders and their power valves are shown in Figure 2. The sequence of these cylinders is also shown in Figure 2 plotted on the circle used to illustrate a sequential circuit. As you can see, this is a good illustration of a sequential circuit because these cylinders must always operate in this order. We do not want the punch to extend until the part is clamped. We do not want the clamp to retract before the punch is out of the part, and so forth. In fact, the function of this

84 Sequential Circuit Design

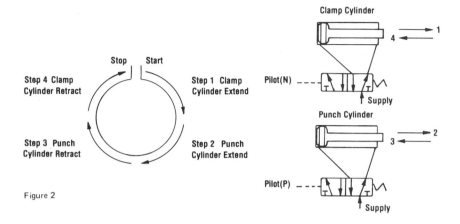

Figure 2

sequence can be very simply designed. Most of the reasons for the design method that will be described are involved in the safety, reliability, and repairability aspects of the circuit. In other words, to make it go is simple; the more complex problem is to prevent actions from occurring when they should not.

Figure 3 shows the inputs to the control circuit. For this illustration we have a limit valve to detect the completion of each of the steps and a single push button to start the automatic cycle. Although the circle diagram in Figure 3 shows four steps, the sequence of operation, when it is written, will include all changes to the input, circuit, or output conditions.

Figure 4 shows the input and output symbols for the circuit. As a general rule, all input signals are shown on the left, control hardware in the middle, and output signals to power devices on the right. Also, where possible, the circuit is drawn as it progresses from top to bottom. The triangles have been added to this drawing to indicate the action which will actuate each limit valve. Use this diagram to follow the sequence of operation.

1. Operator presses PB-1 (input change).
   a. Clamp cylinder extends (circuit or output change).
2. Clamp cylinder releases LV-D.
   a. No change.
3. Clamp cylinder actuates LV-A.
   a. Punch cylinder extends.
4. Punch cylinder releases LV-C.
   a. No change.
5. Punch cylinder actuates LV-B.

Sequential Circuit Design Method   85

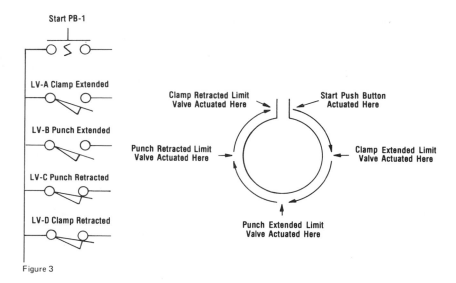

Figure 3

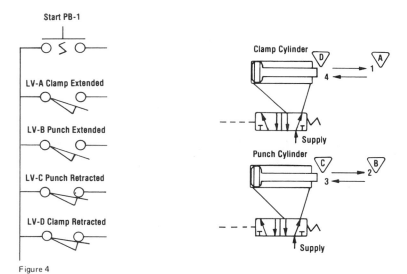

Figure 4

86 Sequential Circuit Design

        a. Punch cylinder retracts.
6. Punch cylinder releases LV-B.
        a. No change.
7. Punch cylinder actuates LV-C.
        a. Clamp cylinder retracts.
8. Clamp cylinder releases LV-A.
        a. No change.
9. Clamp cylinder actuates LV-D.
        a. Circuit reset.
        b. Cycle complete.

Now that we have accumulated sufficient information, we are ready to begin the design of the circuit. As you will recall, the intent is to build within the circuit an entire history of one sequence. The method we will use is a modification of a full ring counter with accumulating code. A ring counter keeps track of the program (or sequence) by memorizing each input signal as it occurs. Accumulating code means that none of these memories (normally flip flops) will be reset until the cycle is complete. The modification to this is simply that we will sort out the signals which need to be memorized and those which do not. Those which need to be memorized will signal a flip flop. Those which do not need to be memorized will signal one input of an And element. This modification is extremely cost effective in most types of hardware since the And element is a less complicated device.

In addition, the signal which need not be maintained will often be present at the end of the cycle, when all of the memories are to be reset. This means that additional elements would be required to remove these signals if all were memorized. So, Rule 1 of this method is as follows: Any signal which is not maintained from the time it is used to the end of the cycle will be memorized with a flip flop.

Figure 5 shows the timing of the start push-button input signal. Although we are not sure exactly when the operator will release this push button, there is no intention that the operator will maintain this button throughout the cycle. Thus, the statement made in rule 1 applies to the push-button input signal.

Figure 6 shows the portion of the circuit developed thus far. The operator presses the push button. This shifts the flip flop so that the signal at step 1 is on. This in turn shifts the power valve and extends the clamp cylinder. When the operator releases the start push button, the clamp cylinder will remain extended. This completes the requirements for step 1.

Step 2 is initiated when the clamp cylinder actuates limit valve A. Again, we must examine the signal from limit valve A to determine if it will

Sequential Circuit Design Method 87

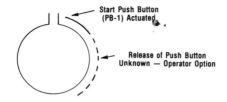

Figure 5

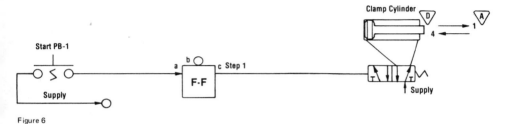

Figure 6

remain on to the end of the cycle. Figure 7 shows that the signal from limit valve A is lost at the beginning of step 4. The signal from limit valve A is, therefore, not maintained to the end of the cycle and rule 1 will again apply.

Figure 8 shows the portion of the circuit now designed. When the clamp cylinder has extended, limit valve A is actuated. This shifts the second flip flop, and step 2 goes on and extends the punch cylinder.

When the punch cylinder has fully extended, it will actuate limit valve B. Figure 9 shows the amount of time the signal from limit valve B will be on. Again, rule 1 will apply. Figure 10 shows the addition of the third flip flop to our circuit. This brings us to rule 2.: The memories (flip flops in this case) are not reset until the end of the cycle. Ouput signals can be removed by using Not elements.

The first part of this rule means that we cannot reset the second flip flop to return the punch cylinder. This relates to the fact that we are building a history of the sequence within the circuit. To reset this flip flop (which, incidentally, would be difficult because we have a signal from limit valve A present) would erase part of this history. In effect, the circuit would be saying that the clamp cylinder has not extended.

The second part of the rule gives you an alternate method of removing this output. Figure 11 shows a Not element added to this circuit which will remove the output signal to the punch cylinder power valve. When the

88 Sequential Circuit Design

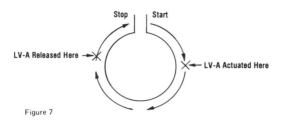

Figure 7

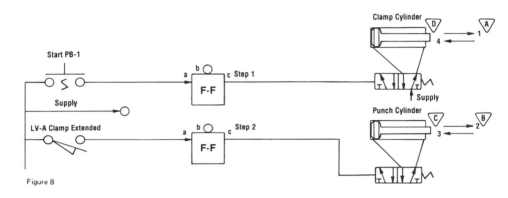

Figure 8

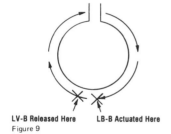

LV-B Released Here   LB-B Actuated Here
Figure 9

signal is present at step 2 only, the output of the Not element will be on (b · $\bar{a}$ = c). When the step 3 signal comes on, the output of the Not element will be off (b · a = $\bar{c}$).

Now the punch cylinder has extended and retracted. Limit valve C will start the next step in the sequence. Limit vlave C will be maintained actuated since the punch cylinder has completed its action and returned. The signal from limit valve C is, therefore, unlike limit valves A and B in that it will remain on to the end of the cycle. This illustrates Rule 3: If an input signal is maintained from the time it is used to the end of the cycle, it should

Sequential Circuit Design Method   89

be connected to one input of an And element. The second input to the And element is connected to the output of the previous step.

Figure 12 shows the And element placed in this circuit. The output of the And (called step 4) indicates that: (1) the punch cylinder is retracting (step 3) *and* (2) the punch cylinder-retracted limit valve LV-C is now actuated. The And element, therefore, prevents limit valve C from having any influence on the circuit until it is time for it to do so.

The step 4 function is to return the clamp cylinder. This is done by adding a Not element to the circuit (see Figure 13). Not element 6 removes

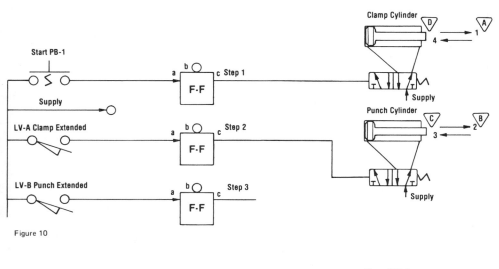

Figure 10

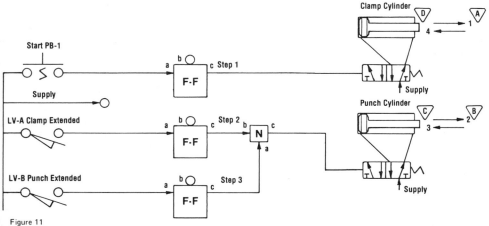

Figure 11

90 Sequential Circuit Design

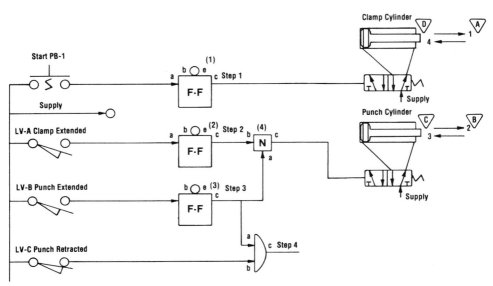

Figure 12

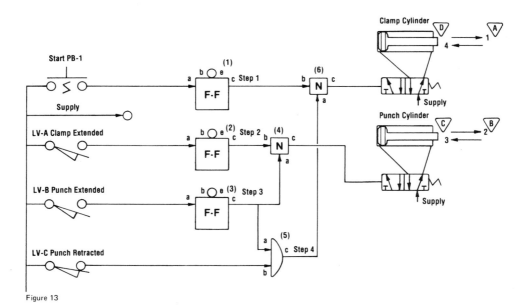

Figure 13

the signal to the clamp cylinder in the same manner as Not 4 did to the punch cylinder. This causes the clamp cylinder to retract.

The final phase of this circuit is to be sure the clamp cylinder is retracted and then to reset the circuit so that it will be ready to begin a new cycle.

Figure 14 shows a new And element (And 7) connecting step 4 (the signal to retract the clamp cylinder) with the clamp-retracted limit valve (LV-D). The output of this And element becomes the signal which will reset the circuit. Figure 15 shows this reset signal connected to the first flip flop. This brings us to Rule 4: The flip flops in the circuit are to be reset in the same order as they are set.

Figure 16 shows the flip flop reset in this manner. Notice that the reset *output* of flip flop 1 (the port marked d) is connected to the reset *input* of flip flop 2 (marked f) and that the reset *output* of 2 is connected to the reset *input* of flip flop 3. So, the circuit will reset in basically the same order it was set. First flip flop 1, then flip flop 2, then flip flop 3, then And 5 (because step 3 signal is now off), then And 7 (because step 4 signal is now off).

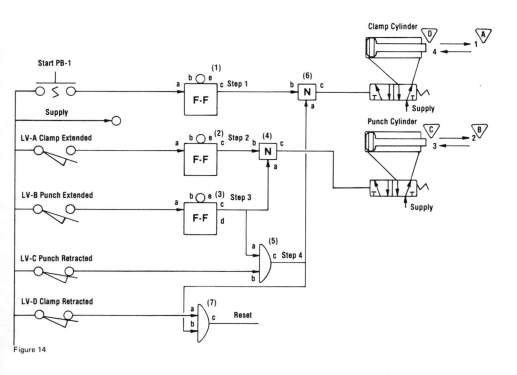

Figure 14

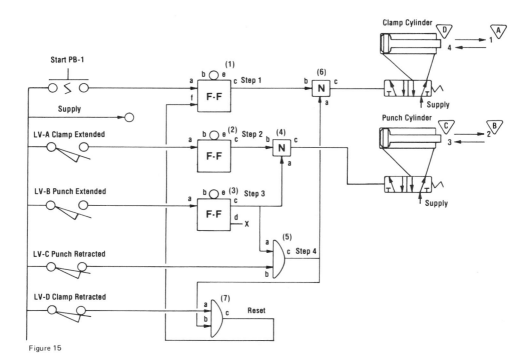

Figure 15

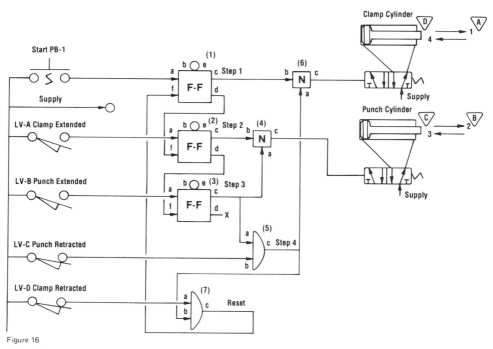

Figure 16

Sequential Circuit Design Method   93

This completes the design of the basic circuit for this sequential circuit. Before we leave this example, we should review this design to see if it complies with the good design practices described in the beginning of this chapter.

Design Evaluation

Function: A review of the sequence of operation and the completed circuit will show that the required function as stated has been accomplished. This is also a good time to check closely points 2, 4, 6, and 8 of the unwritten sequence. Be sure there is no change as requested.

Safety: This is a tough one. Even in a machine as simple as this, there are many potentially unsafe conditions. Here are some of them.

1. *The machine start.* Is the machine properly guarded so that the operator cannot start the cycle with the other hand in the machine? Perhaps a two-hand anti-tie-down should be added to the circuit for safety purposes. This safety problem will often be given proper consideration because of the emphasis placed by OSHA in this area. Still other safety problems will not be as easy to recognize.

2. *Inadvertent actuation of a limit valve.* There is a good example of this in the circuit just designed. Suppose the operator in loading or unloading this machine inadvertently actuates limit valve A. This limit valve normally indicates that the clamp cylinder is holding the part and that the punch cylinder should extend. Will it? Notice that in our circuit drawing the reset output port of flip flop 1 is holding pressure on the reset *input* port of flip flop 2. If limit valve A is actuated pressurizing the set port of flip flop 2, the flip flop will not shift because the reset signal to flip flop 2 is still present. Remember that the flip flop will remain in the position first shifted if both signals are present and of equal value (pressure). If you further trace the circuit, you will find that in the "at rest" position no limit valve signal will cause the circuit to change or the cylinders to move.

3. *Malfunctioning input devices.* There are several examples of safety problems that could result from malfunctioning input devices in this circuit alone. First would be a jammed start button, causing a repeat cycle. Will this happen? Notice that again the flip flop, in this case flip flop 1, works to our advantage. If the start signal is still on when the reset signal arrives at flip flop 1, it will not reset. If flip flop 1 cannot reset, the circuit cannot reset. If the circuit cannot reset, it cannot start over. The second condition would be a jammed clamp-extended limit valve (LV-A). If this occurs, the punch cylinder could advance before the clamp function was complete and throw the part.

Tracing the circuit will again reveal that a malfunction of limit valve A in the On condition would prevent the circuit from resetting on the pre-

94 Sequential Circuit Design

vious cycle. Flip flop 2 would not reset. The balance of the circuit would not reset, and because of this, a new cycle could not be initiated.

The third potential problem is a malfunctioning of limit valve C in the passing condition. This could cause the clamp cylinder to retract early, perhaps while the punch was still in the part. In this case, there are no provisions for this failure automatically provided by the circuit design method we used. If this fault, after consideration, is determined to be a potential safety problem, additional circuitry must be added to prevent its occurrence. Figure 17 shows the addition of a simple interlock which can be used to detect a malfunction of limit valve C or D. The output of this And element could be used to activate the emergency stop system for this circuit—a subject which we will cover in more detail later.

These are the safety aspects of the machine and, as you can see, as circuits become more complicated, the sheer number of safety items that should be considered can be staggering. The method of design used to cre-

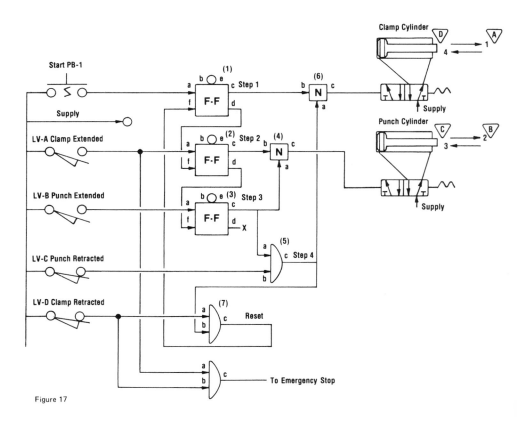

Figure 17

ate the sequential circuit can be a tremendous help to us in safety if it will consistently interlock most of these unsafe conditions and allow us to concentrate on the few it does not.

Reliability: First, consider the components. By producing a design which uses standard functions, you have the widest possible selection of brands. This allows you to select the hardware that will deliver the maximum reliability on this application. Second, consider the design itself. Several things will decrease the overall reliability of a circuit. Pulse (momentary) signals and unnecessary adjustments are two of the most common reliability problems. The circuit should be examined for these characteristics, and they should be eliminated wherever possible. A good design method will eliminate most of these potential problems. During the life of this machine the control circuit will, in all likelihood, experience various operating pressures, tubing leaks, and temperatures, sluggish components, and many other variables. The measure of reliability in a control circuit is how long it can continue to operate in spite of these variables.

Repairability: Now the machine finally stops. What can we do to get it started again? We have already contributed much to this effort by eliminating pulses and unnecessary adjustments from the circuit. By eliminating pulses, for example, we now have a set of maintained signals we can use to analyze the problem. Going one step further, we can provide visual indicators and a simple troubleshooting chart to help determine the nature of the problem without analyzing the drawing of the circuit. Figure 18 shows the addition of an indicator at each phase in the circuit, and Figure 19 shows a troubleshooting chart which would be used with these five indicators.

Cost: There are two ways to reduce the price of the circuit. The first is to compromise some of the features or functions of the circuit itself. In Figure 20 we reduced the circuit by connecting the punch-retracted and the clamp-retracted limits in series with the stage 3 and each other. What we gained was a reduction of the circuit by two elements. What we have lost is:

1. Some ease in troubleshooting: It will be more difficult to test limit valve C and D because they do not have a maintained supply.
2. One more tubing connection is required from the control panel out to the machine.
3. Perhaps some cycling speed, depending on the speed of the machine and the length of connecting lines.

This is an example of a reduction of circuit by reduction of features. A reduction by redefinition of the function is also occasionally possible. In this

96 Sequential Circuit Design

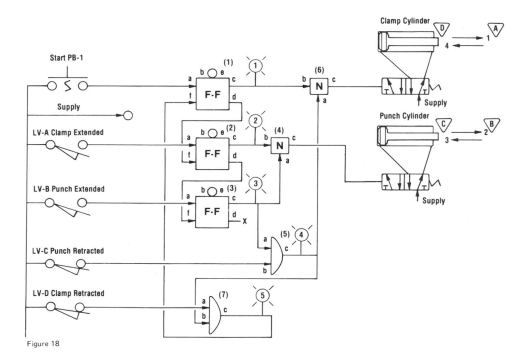

Figure 18

circuit, for example, we could decide not to monitor the retract stroke of the clamp cylinder. The removal of the LV-D limit valve and the associated elements is shown in Figure 21.

The second method to reduce cost is in shopping for control hardware—balancing the need for reliability and the reliability offered by various brands and their price. Using standard functions such as the elements called out by this design method gives you the widest latitude possible in your selection of components.

There are some good guidelines to follow when working on circuit cost reductions. First, in the area of reducing features or functions, be sure in each case you can identify the feature or function you have lost as a result of any change. Except in the case of an absolute redundancy in the circuit, there will be some loss of feature or function. Identify this carefully, and remember that you buy the hardware only once, but you live with the control as long as the machine lasts. Under no circumstance can a reduction to a safety feature be justified.

In hardware selection, it is generally a good policy to go with one brand for all of the components in a system. Although manufacturers de-

Modifications to Basic Design 97

sign their components to work well with each other, they do not necessarily design them to be compatible with those of other manufacturers. There is no standardization in this area to date. Also, the same rule applies to decisions on quality, as mentioned before. You only buy it once, but you must use the control for a long period of time. Small savings in the initial price can be used up many times if the quality is less than adequate.

MODIFICATIONS TO BASIC DESIGN

Some modifications or additions are normally required to every basic design. An emergency stop system or a manual control system are typical examples of additions that often must be incorporated. These additions are combinational circuits. The emergency stop push button, for example, can be operated at any time. Manual controls, in many cases, can be operated in any sequence. For this reason they cannot be designed as a part of the sequential circuit, but must be considered separately.

Troubleshooting Chart

Instructions: Select the line below which represents the condition of the troubleshooting indicators, follow the instructions for that line in order. (indicates condition that should be found)

Indicators	Actions
① ② ③ ④ ⑤	1. Check air supply. 2. Check PB-1 for proper function. 3. Replace Flip Flop 1.
⊗ ② ③ ④ ⑤	1. Check clamp cylinder (extended) if not problem is in cylinder or valve. 2. Check limit valve A (output on). 3. Replace Flip Flop 2.
⊗ ⊗ ③ ④ ⑤	1. Check punch cylinder (extended). If not problem is in cylinder or valve. 2. Check limit valve B (output on). 3. Replace Flip Flop 3.
⊗ ⊗ ⊗ ④ ⑤	1. Check punch cylinder (retracted). If not problem is in cylinder or valve. 2. Check limit valve C (output on). 3. Replace "AND" 5.
⊗ ⊗ ⊗ ⊗ ⑤	1. Check clamp cylinder (retracted). If not problem is in cylinder or valve. 2. Check limit valve D (output on). 3. Replace "AND" 7.
⊗ ⊗ ⊗ ⊗ ⊗	1. Check PB-1 (output off). 2. Replace Flip Flop 1.
① ⊗ ⊗ ⊗ ⊗	1. Check limit valve A (output off). 2. Replace Flip Flop 2.
① ② ⊗ ⊗ ⊗	1. Check limit valve B (output off). 2. Replace Flip Flop 3.

Figure 19

98 Sequential Circuit Design

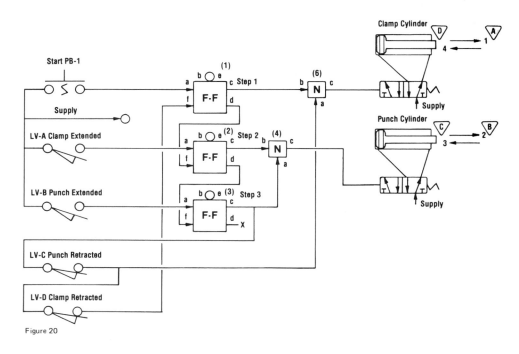

Figure 20

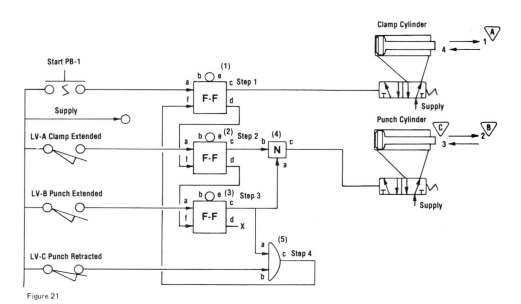

Figure 21

Modifications to Basic Design 99

Emergency Stop Design

Emergency stop controls can be a study in themselves. Practically every machine will have its own requirements for the actions that should take place when the operator pushes "that red button." For this reason there has been no national standard developed for emergency stop functions, and companies who have experimented with corporate standards have found it very frustrating. They soon find that in applying their standard method to various machines they are as often wrong as right.

Since we cannot give you a standard method to design all emergency stop sytems, the next best solution is to list and describe some of the options you have and try to describe where they may be useful. You must then study the requirements of each machine and determine which one of these options or combination of these options best suits the application.

Before you begin your design, you should be able to describe the actions that will take place when the emergency stop push button is actuated. Second, you should be able to describe the actions that will take place when the machine is restarted. The options for emergency stop, although numerous, fall into four basic types. They are:

1. Hold
2. Relax
3. Abort
4. Return in sequence

Hold

Hold emergency stop systems are used to stop and hold the cylinders in position when the emergency stop push button is actuated. Figure 22 shows the clamp and punch circuit with a simple hold emergency stop option. We have replaced the two-position four-way valves with three-position (all ports blocked) four-way valves. When no pilot signals are present, these valves will return to the center position and stop the cylinder. The emergency stop push button is normally passing and feeds the supply to the circuit. When the emergency stop push button is actuated, this supply and subsequently all output signals from the circuit are removed. When the emergency stop push button (here shown as a push-pull maintained button) is reset to the passing mode, the cycle will resume from the point where it was interrupted. Although shown here with air cylinders, this type of emergency stop is more commonly used with hydraulic cylinders. Since hydraulic fluid is not compressible, they will have much better holding characteristics than would air cylinders under the same circumstances. Notice also that signals from steps 3 and 4 are used as the return pilot sig-

100 Sequential Circuit Design

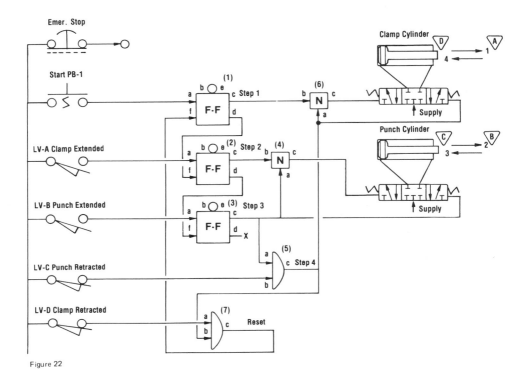

Figure 22

nals to the three-position valves. With this circuit the valves would center when the machine is idle (not uncommon in hydraulic power systems).

The circuit in Figure 23 will provide the same emergency stop function as Figure 22, except that in the idle condition both cylinders will be held retracted under pressure (more characteristics of pneumatic systems). Obviously, the hold emergency stop system cannot be used in cases where gravity or other forces would cause the cylinders to move.

Relax

A relax emergency stop is often used on small pneumatic machines and fixtures. This system removes the pressure from both sides of the piston in the air cylinder so that it can be moved freely. On smaller fixtures the operator can then "unjam" the machine by moving the cylinders by hand.

Figure 24 shows how this system could be designed using three-position valves. The center position on this valve opens both cylinder ports to exhaust. The hydraulic equivalent would connect both cylinder ports to tank.

Modifications to Basic Design 101

An alternate method often used for pneumatic cylinders is shown in Figure 25. This emergency stop operates a master dump valve which removes all pressure to the power cylinder and valves. The hydraulic version of this is to cut the power to the pump motor.

Although the relax emergency stop is simple to accomplish and in some cases very useful, again you must be sure that gravity and other forces will not cause the cylinders to move. One other problem occurs when this emergency stop method is used with air cylinders. Control of the cylinder speeds, normally provided by flow controls, is lost when all pressure to the cylinders is removed and reapplied. To prevent "slamming" of these cylinders when the system is restarted, it may be necessary to install a two-stage restart system similar to those shown in Figure 26. This allows a slow pressure build-up before the actual restart occurs.

Abort

Abort emergency stop systems immediately return all cylinders to their original positions and reset the circuit. Figure 27 shows this emergency stop option added to the clamp and punch circuit. With Or element 8 added

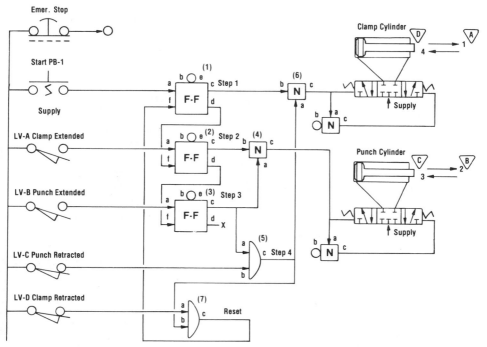

Figure 23

102 Sequential Circuit Design

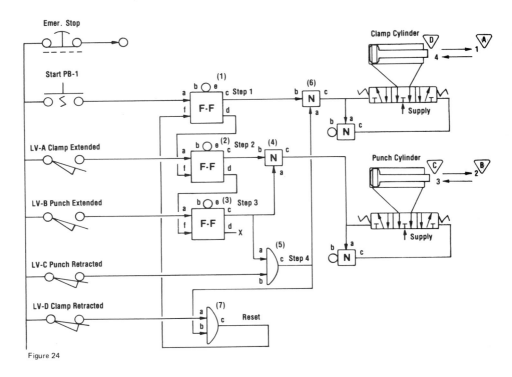

Figure 24

to this circuit, flip flop 1 can be reset by the emergency stop push button at any time. Once flip flop 1 is reset, the balance of the flip flops will reset as fast as the set inputs are removed.

In many cases, this can be an adequate form of emergency stop; however, be sure to analyze the sequence carefully. Our circuit is a good example of the problems that might be encountered using this method. In one circumstance (clamp extended, punch extending), this emergency stop will actually cause the clamp to retract (reset of flip flop 1) before causing the punch to reverse (reset of flip flop 2). Flip flop 2 is prevented from resetting until limit valve 2 is released. Depending on the machine itself, this may or may not be the best emergency stop system to use.

Return in Sequence

Often, after considering several possibilities, the best option for emergency stop may be to return all of the cylinders to their original position, but in the sequence they would normally follow. The clamp and punch machine could be an example of this. Figure 28 shows how this emergency stop can

Modifications to Basic Design   103

be incorporated into our circuit. By connecting a maintained emergency stop signal to Or element 8, the emergency stop signal starts the retract portion of the automatic sequence. It will first cut off the signal to the punch cylinder. Also, it will combine with the punch-retracted signal at And 5 to retract the clamp. Finally, when the clamp-retracted signal is on, it will reset the circuit. The emergency stop button will then continue to inhibit a new cycle start until it is reset.

These are some of the more common types of emergency stop sequences. There will be occasions when combinations of these sequences will be used. In still other cases it may be necessary for the emergency stop system to provide an entirely unique sequence of operation when actuated. In some cases you may find that the design of the emergency stop will involve as much consideration as the automatic sequence and may require some changes to the basic hardware for greatest efficiency. For this rea-

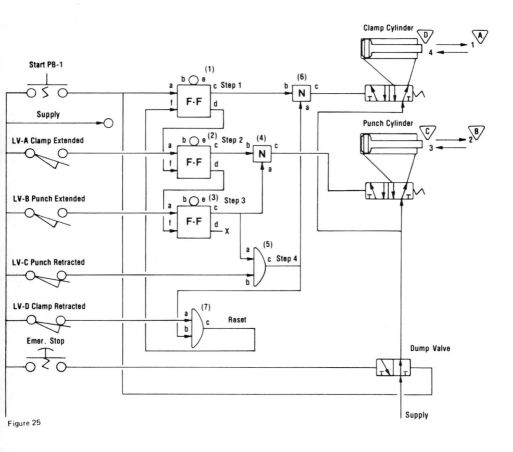

Figure 25

104 Sequential Circuit Design

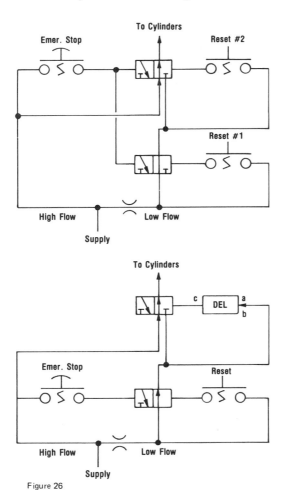

Figure 26

son, we recommend that you consider the emergency stop sequence very carefully each time you have completed the basic circuit.

Manual Controls

Incorporating manual control into an automatic system is quite simple. Interlocking the manual controls so that they in themselves do not create a problem can make this more complex.

Modifications to Basic Design 105

Figure 29 shows an addition of an automatic/manual selector valve, two manual selectors, and two Or elements for independent manual operation of the two cylinders. The auto. side of the auto./manual selector is connected in series with the start push button only. Because of the interlocks described previously, the other input signals to the automatic portion of this circuit will have no effect, even though they do change as a result of the manual movement of the cylinders. Additional manual interlocks can be added to prevent undesirable conditions.

Figure 30 shows the addition of an And element to prevent the punch cylinder from extending prior to the clamp. Notice that this interlock only ensures the sequence of extension of the two cylinders. They can both be retracted by selecting only the clamp manual selector off, or the auto./manual selector to the off or auto. position. Thus, there may be a need for additional interlocks to prevent this action.

Figure 31 shows a "full-blown" manual sequence section added to this circuit. This manual sequence section interlocks the following:

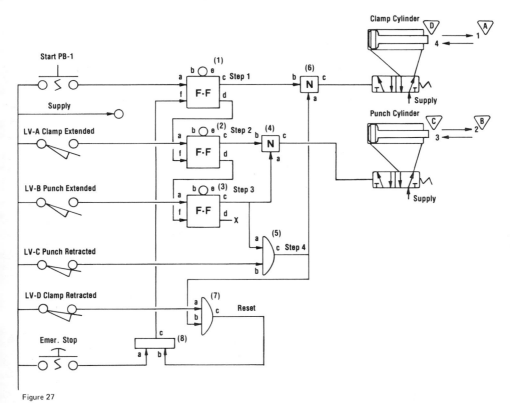

Figure 27

106 Sequential Circuit Design

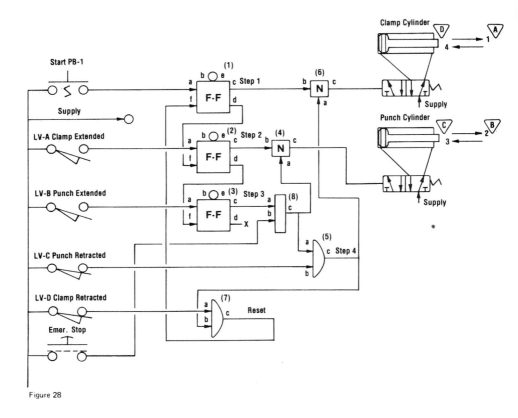

Figure 28

1. Manual push buttons and auto. start button with auto.-off-manual selector (series connection).
2. Prevents the auto.-off-manual selector from having a direct effect on the cylinders through the use of push buttons and flip flops 11 and 13.
3. Prevents the punch from extending until the clamp is extended (And 12).
4. Prevents the clamp from retracting until the punch is retracted (And 10).
5. Prevents the auto. sequence from starting until the manual circuit has been cleared (Or 14 and Not 15).

This brings us to another important part of manual control design—cost-cutting measures for manual controls. Here is where a thorough analysis of the machine by the designer can pay real dividends. The questions

Modifications to Basic Design 107

that you must analyze are: Are these manual controls really necessary for the proper operation of this machine? What do they do? Who uses them, and what do they use them for?

Many machines have far too many manual controls. These controls were included only because someone thought they "might be useful sometime," or out of habit. Unless you can describe the specific need for these controls, then you may find that they are not necessary.

Second, who are you "giving" these controls to? In Figure 31, we are giving these controls to the operator by placing them on the operator control panel. Perhaps the operator is not the person who should have these controls. If only the set-up or repair people need these controls, then don't put them on the operator control panel. Put them in the control en-

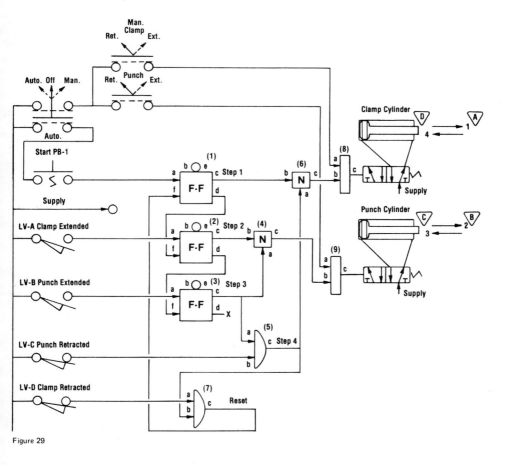

Figure 29

108　Sequential Circuit Design

closure or at some other location. By removing them from the operator control panel and the day-to-day possibility of use, you may have removed much of the need for interlocking these controls. The manual controls could be, in this case, as simple as a manual override on the four-way power valves.

Figure 32 illustrates a simple addition to this circuit which will allow the cylinders to be manually or automatically operated. This addition could be used during machine build, set-up, or troubleshooting, and then *removed* so that it cannot be used by unauthorized personnel.

Analyzing the machine requirements carefully and then "giving" the manual controls only to those who need them and are properly instructed on how to use them can greatly reduce the complexity and cost of the manual control system.

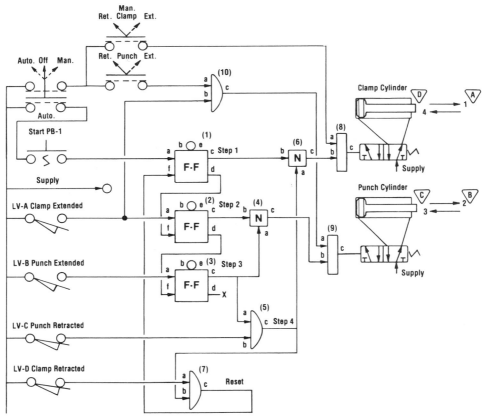

Figure 30

Modifications to Basic Design 109

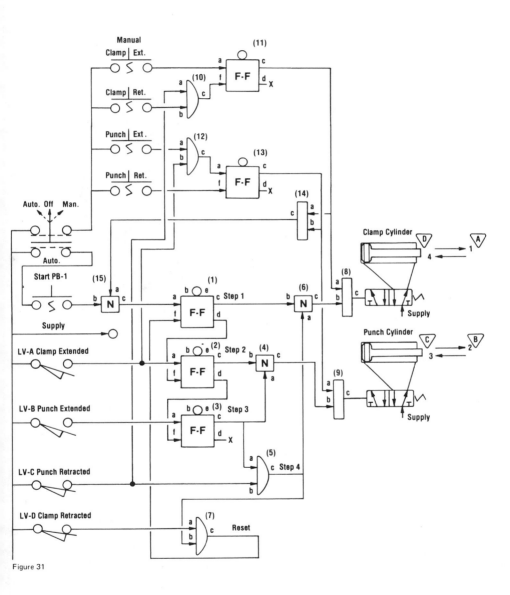

Figure 31

110 Sequential Circuit Design

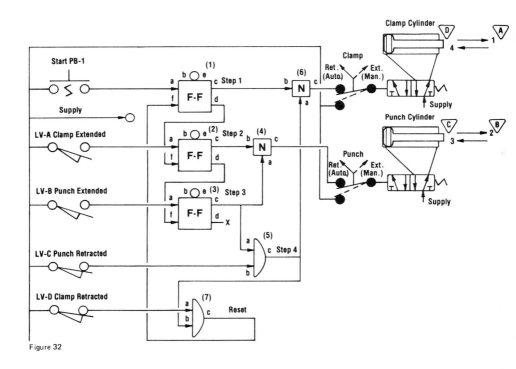

Figure 32

9
VARIATIONS OF SEQUENTIAL CIRCUITS

INTRODUCTION

In Chapter 8 we studied one basic circuit in great detail. In the chapters that follow we will show as many variations of circuits as possible. By studying these circuits and their descriptions, you will be able to learn additional rules that will be helpful in designing circuits with similar requirements.

We have all studied methods in the past that work great on one or two textbook examples, but when we try to use the method in real-life situations we fail. This is either because the method was faulty in the first place (which we will believe forever), or that not enough explanation and examples were given to help us prepare for the variety of situations we will encounter. By showing a variety of applications we hope to:

1. Show that the method described in the last chapter will work on all sequential circuits.
2. Show any additional rules and tips that apply to specific circumstances.
3. Provide additional experience in a wide variety of control applications.

For your review before you start, here are the design rules given for sequential circuits in the previous chapter.

1. Signals which are not maintained from the point they are used to the end of the cycle are memorized with a flip flop element.
2. These flip flop elements are not reset until the end of the cycle.

3. Not elements are used to remove output signals when they are no longer required.
4. Signals which are maintained from the time they are used to the end of the cycle are connected to one input of an And element. The other input to this And element is provided by the previous stage.
5. The reset signal is connected to the reset input of the first flip flop. The balance of the flip flops are reset in the order they are set (reset output of the first to reset input of the second, and so forth).

These are the basic rules, and in the examples that follow we will describe only the variations or modifications that need be applied in these "real-life" applications.

EXPANSION

Figure 1 shows a circuit designed for three cylinders and six steps. Following the method previously described, the circuit has expanded by one flip flop (addition of a nonmaintained limit), one And (addition of a maintained limit), and one Not (an additional output that must be removed).

All statements regarding function, safety, reliability, troubleshooting, and cost-reduction techniques remain valid. All statements regarding emergency stop and manual options also remain valid. Thus, the method can be used on circuits of any number of power devices and steps.

MIXING

Figure 2 may at first appear to be quite different from Figure 1. On closer examination, however, you will find that the sequence is merely scrambled from the sequences previously shown. Here are the changes in this circuit and their effect on the design:

1. A double pilot operated four-way valve operating cylinder number 1. This has no effect on the basic design method, since pilot signals are available for each action in the sequence. The only real change this may make in the design of the circuit would be in the emergency stop and manual controls, depending on the methods chosen.
2. Cylinder 2 is normally extended. This is done by reversing the connections to the cylinder from the power valve. Although it adds to the confusion in following the sequence, it does not require any change in the method used to design this circuit. If this becomes troublesome to you in the design of a circuit, it is possi-

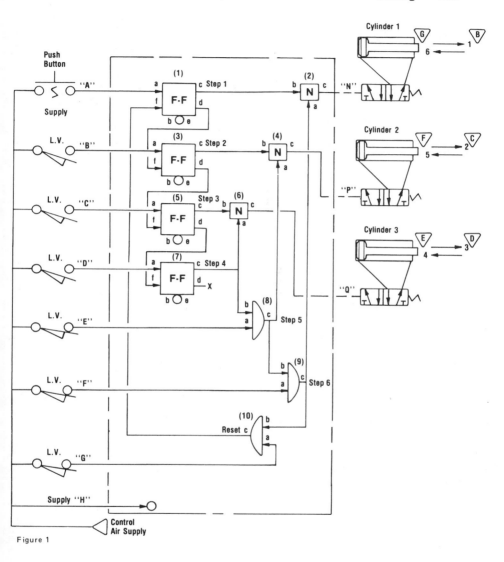

Figure 1

ble to design the circuit with all cylinders normally retracted. Then, reverse the connections to the cylinder and the position of the two limit valves on your final drawing.

3. In our first two examples, the cylinders all extended then returned in their reverse order—nice and neat. This example shows a sequence more on the order of what you will experience in actual

114 Variations of Sequential Circuits

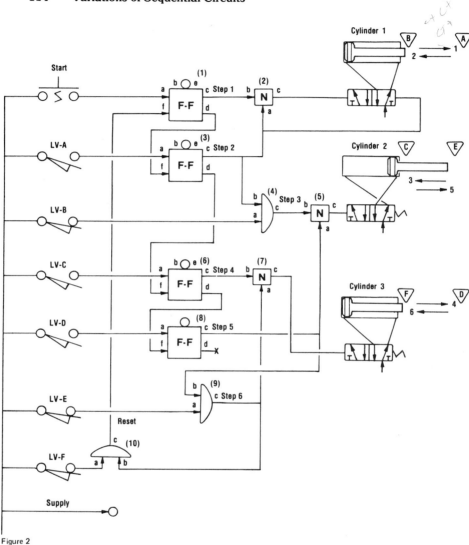

Figure 2

applications—scrambled. In previous examples, the flip flops for nonmaintained signals formed the first stages of the circuit. Here we find that a maintained signal comes into play earlier in the sequence (limit valve B). Nevertheless, the rules for maintained and nonmaintained signals still apply and the And element was used to develop the step 3 signal. Cylinders 2 and 3 also perform a different sequence than those previously shown, even disregarding the fact that cylinder 2 is normally extended. This change in the sequence is made merely by the connections made from the step outputs to the Not elements and the order in which the limit valves come into play. Following these examples, then, it must be assumed that this method of design will produce a valid circuit for any number of actions and in any sequence.

4. One other thing, before we leave this example: Notice that cylinder 1 extends and retracts in the first two steps of the sequence. When cylinder 1 has retracted (end of step 2), the signals to the circuit are identical to those present in the beginning. This demonstrates the futility of attempting to design sequential circuits with combinational logic methods. No matter how much mapping, diagramming, analysis, or manipulation is done, the fact remains that the memory is essential to this circuit. It must be included to distinguish between the beginning of the cycle and the end of step 2.

CONTRACTION

Figure 3 shows an application which involves six cylinders. You might at first expect that the circuit for this machine would contain twice the number of components as were required for the three-cylinder circuits. As you can see, this is not the case. In this circuit, cylinders 1 and 2 and cylinders 5 and 6 are operated from the same power valve. These cylinder sets are incapable of independent action. The only difference between these two cylinders in a set and a single cylinder is the use of two limit valves, one for each cylinder. In some cases this may not be necessary if they are mechanically linked. In any event, even if two limit valves are used they can often be connected in series (that is, A^1 and A^2) and represent one input to the control system.

This circuit also has examples of conditions where two or more cylinders, although capable of independent action, operate at the same time. An example of this is step 2. Here cylinder 3, although capable of independent action, operates in unison with cylinders 5 and 6 on the extend stroke and with 1, 2, and 4 on the retract stroke. All of these combinations of actions tend to simplify rather than complicate the control system. For this reason, you will find that the number of circuit components required for the basic circuit will follow much more closely the number of steps in the sequence of operation than it does the number of power devices.

Other provisions in the circuit may require that the circuit "see"

116   Variations of Sequential Circuits

some of these limit valves independently of others. Manual control interlocks could be an example of this. In this case, And elements would be used in the circuit to combine the limit valve outputs rather than the series connection shown. Whether the limits are in series or combined with And elements, be sure to select the flip flop or And element for each stage based on the combination of all the limit valves. Step 4, for example, would have been an And element except for limit valve C^3. Here again, no changes in the rules of design.

DOUBLE SEQUENCE

Figures 4 and 5 are of different but quite similar sequences. Figure 4 is a pretty straightforward three-cylinder sequence of yet another combination of steps. Figure 5 is the same as Figure 4 except that cylinder 3 has been eliminated and cylinder 1 performs two operations. The reason that we have shown them both is to show the differences between the circuits. The first is the addition of Or 3 so that both Not 9 and Not 2 can control cylinder 1. The second is the addition of flip flop 6. Notice that we did not remove And 5. That means that the logic required to create step 3 is the combination of an And element and a flip flop element. The signal from limit valve C when the number 1 cylinder is used twice is unique. If you were to describe the action of limit valve C input after it starts step 3, you would say that it will be lost before the end of the cycle (at the beginning of step 4). This would call for a flip flop in position 3.

You would also say that limit C is on (maintained) at the end of the cycle (beginning at the end of step 5—normally the requirement for an And element). So, what you really have is a requirement for both in step 3, and both are included—the flip flop to prevent the loss of the step 3 signal and the And element which allows the circuit to reset flip flop 6 at the end of the cycle. The circuit shown is also used when the condition of the limit valve is "uncertain" throughout the rest of the cycle or at the end of the cycle. For example, a flywheel may or may not come to rest on a limit valve at the end of a cycle, or material on a conveyor should produce a maintained signal at a limit valve, but someone might remove it by hand. These types of circumstances can be overcome by using this combination of the And and the flip flop connected as shown in Figure 5.

Some sequential design methods recommend the use of this combination or other similar combinations for all steps in the sequence. Indeed, a functioning circuit can be designed in this manner. However, as you can see, this would increase the complexity and cost of the circuit. In addition, using the And/flip flop combination for every step actually removes some interlocks which were valuable (the functional checking of inputs con-

Double Sequence   117

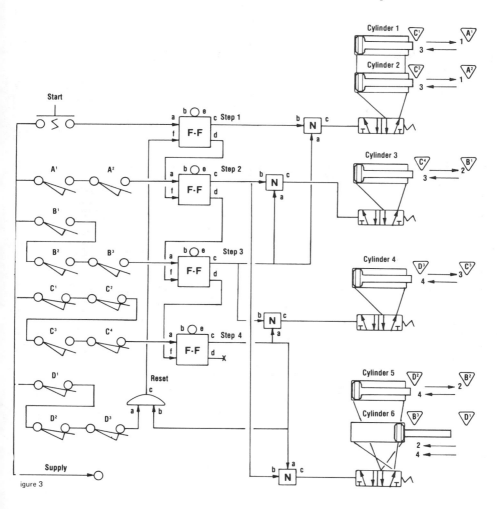

Figure 3

nected to flip flops each cycle). So, this combination should normally be used only when we know the signal will be released during the cycle, but actuated and maintained at the end of the cycle or in cases where we cannot determine the condition of the signal at end of cycle.

Before we leave Figure 5, be sure to examine this sequence closely and note the following. At the beginning of the sequence, cylinder 1 extends and actuates limit valve B. Now trace B input. B input actuates the set port of flip flop 4 and flip flop 10. Flip flop 4 will shift—flip flop 10 will not because it is being held reset by flip flop 8. Cylinder 1 retracts. Now we

118 Variations of Sequential Circuits

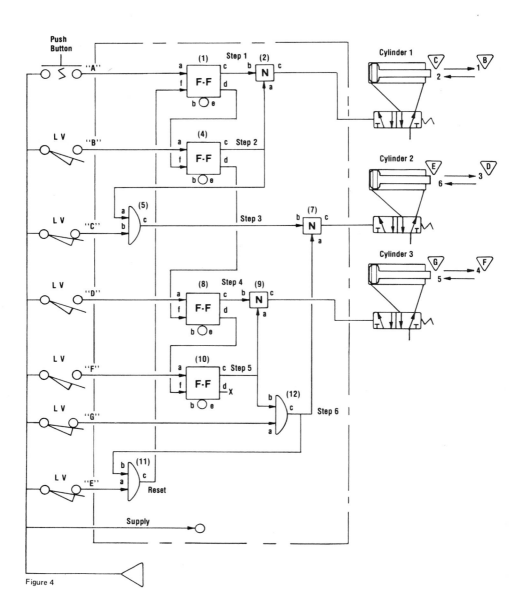

Figure 4

Double Sequence 119

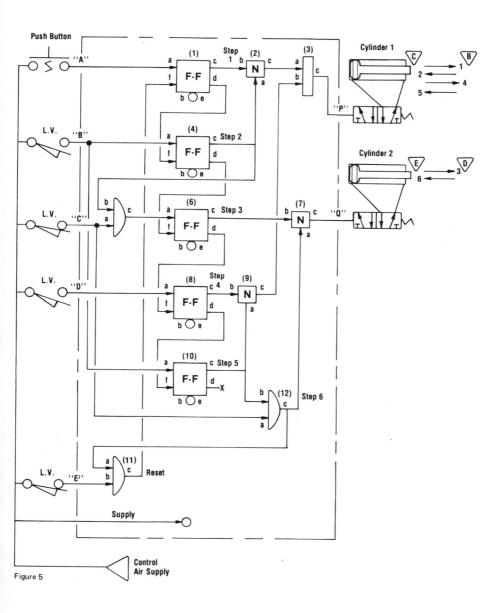

Figure 5

120   Variations of Sequential Circuits

actuate limit C. Signal C is connected to And 5 and And 12. And 5 will shift, but 12 will not because step 5 has not been set. As you follow this to completion, you can see that interlocks provided by the flip flop and And elements work for us in helping the circuit keep track of its own position in the sequence.

CONTINUOUS OPERATION

Designing continuously operating circuits can be easily accomplished. Figure 6 shows the clamp and punch circuit from Chapter 8 converted to continuous operation. The push button has been replaced with a maintained selector switch, and Not 8 was added to the circuit to remove the set signal to flip flop 1 during the reset phase. Note that we are using this only as an illustration. Obviously the application would not be a good application for continuous operation.

More complex circuits are often required in place of the on-off selector. Some of these include start and stop push buttons, and single cycle and automatic (continuous) combinations. Some of these will be shown in

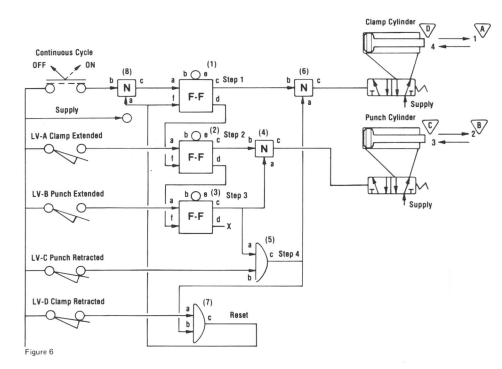

Figure 6

later circuits. Nevertheless, here are some items that should be considered whenever the circuit is to include a continuous operation mode.

The first is safety, since this modification will free the operator. The second is emergency stop. Additional circuitry may be required to cancel the start signal during emergency stop—in this example at Not 8. Restart from emergency stop should be studied carefully. In this example, if the continuous cycle selector were to remain in the on condition, the machine would start as soon as the emergency stop signal was removed. This often leads to the more elaborate push-button starting systems mentioned before.

Finally, monitoring the full sequence of the machine is normally essential. For example, the reduced circuit shown in Figure 20 of Chapter 8, where we chose not to monitor the retract stroke of the clamp cylinder, would not function properly in continuous operation.

DRIVEN SEQUENCE

This type of sequence is different from single cycle or continuous in that for all or a portion of the sequence the start signal must be maintained. This is normally a safety requirement and, in effect, is a form of emergency stop.

Figure 7 shows the clamp and punch circuit with a driven sequence modification through steps 1 and 2. Here it was decided that the danger to the operator existed until step 2 was completed (Ands 8 and 9 were added). Then it was decided that the cylinder should stop if either start push button was released during this portion of the cycle (anti-tie-down function 10 added). Later it was decided that the three-position valves shown would be adequate, but only if they were held in the power-retract position during the idle portion of the cycle. Or elements 11 and 12 were added to maintain the pilot signals to the retract ports of these valves when the machine was in the idle mode (see Figure 8).

ELECTRICAL COMPARISON

Figure 8 also serves to illustrate another important fact. The basic characteristics of a pneumatic circuit are used throughout the design of these circuits. Figure 8 shows enough of the differences between the characteristics of an air and an electrical circuit to be useful as an illustration on this as well.

For example, here we are holding the pilot signal on the three-position valves during the idle period of the cycle. In air circuitry, this is no problem. If, however, this were an electrical control circuit operating

122   Variations of Sequential Circuits

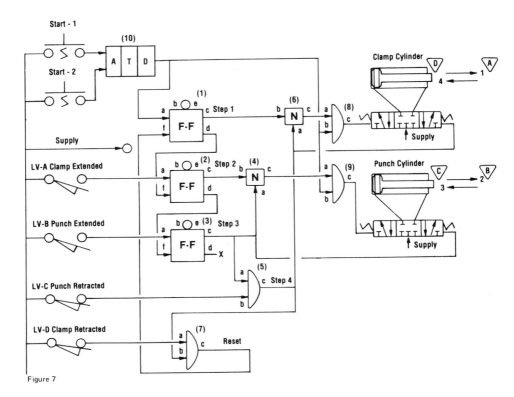

Figure 7

three-position solenoid valves, additional investigation and perhaps change to the circuit would be required. The solenoid itself would consume power and can be destructive to itself when used in this manner. The pilot line, once filled and barring leakage, consumes no power, builds no heat, and generally can be used exactly as shown.

The same characteristics apply to the control components themselves. Again, for example, the equivalent electromechanical device to the flip flop will have entirely different parameters for its use. Depending on the type selected, the following statements could easily be true:

1. It could be inadvisable to connect the reset signals as shown in this illustration because one coil on flip flops 2 and 3 would be energized during idle periods.
2. It could be inadvisible to use the flip flop for a sequence interlock as we have described. The electrical equivalent may be unpredictable (as to the position it will assume) or damage to the component itself may be caused by energizing both coils at the same time.

All of this points out two "don'ts" in circuit design:

1. Don't use the method we describe here to design electrical circuits.
2. Don't convert electrical circuits directly to pneumatic. Many times the circuits designed for electrical controls will either not work properly as pneumatic controls or will contain provisions which are not necessary for pneumatic controls, causing your pneumatic "copy" to be unnecessarily redundant, confusing, and costly.

REPEATERS

Circuits which repeat for a given number of cycles, a period of time, or until they receive a stop signal are often required. Figure 9 shows the clamp and punch circuit modified to repeat until a given number of cycles have been completed. Three elements have been added: Or 10 (which allows flip flop 2 and the balance of the circuit to be reset from two points) and And 9 and Not 8, which function as a redirect circuit. As long as the counter has no

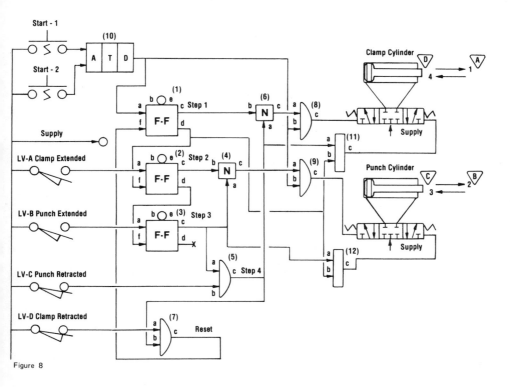

Figure 8

124   Variations of Sequential Circuits

ouput, the signal from And 7 is directed to Or 10, resetting the lower portion of the circuit. When the last count is received and the counter output comes on, the signal from And 7 is stopped by Not 8 and redirected through And 9 to reset the entire circuit. When the circuit is reset, flip flop 1 also resets the counter.

Quite often, repeating circuits will be combined with other nonrepeating functions in the same machine. Chapter 10 has other illustrations of repeating circuits and tips on how to interlock repeating with nonrepeating circuits.

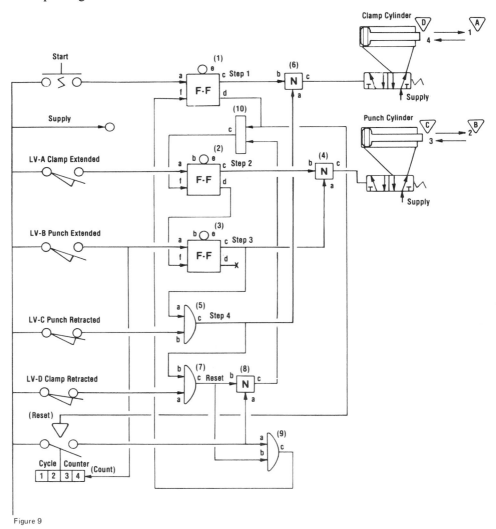

Figure 9

10
INPUT VARIATIONS AND OTHER SPECIAL CIRCUMSTANCES

INTRODUCTION

All of the examples up to now have been of circuits which included limit valves to detect every motion. Wherever possible, limit valves are normally the logical first choice. Combined with a circuit that will double-check the limit valve to be sure it is working properly, the limit valve is the most positive means we have to detect that a movement has taken place. Unfortunately, in real-life situations this is not always possible or practical.

Timers, pressure signals, bleed signals, jet sensing, and other methods all have applications in pneumatic controls. For a design method to be valuable, it must be able to incorporate all of these and more.

This chapter will continue to explore various types of circuits, but in addition we will incorporate a variety of these devices into the circuits showing what, if any, changes are required in the method of circuit design previously shown. In addition, as each is introduced, we will attempt to describe the advantages and disadvantages of the various devices used.

BLEED FUNCTIONS

Bleed functions can directly replace limit valves. The bleed function is nothing more than a two-way normally closed valve. Because of its simplicity, it can sometimes be installed in areas in which the three-way limit valve cannot. Figure 1 shows a two-way bleed function installed in place of the three-way valve (LV-A). The two- to three-way converter (8) converts the bleed signal to a three-way on/off signal. This signal is used to actuate

126   Input Variations and Other Special Circumstances

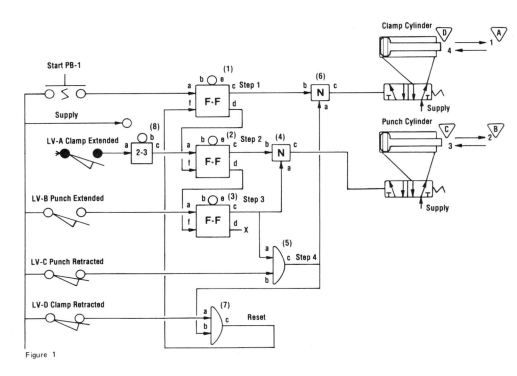

Figure 1

flip flop 2. The design method need not be altered in any way. The two-way (bleed) valve and the two- to three-way converter simply replace the three way valve in the circuit.

Whenever using a bleed valve in any system, be sure to carefully consider the following negative factors:

1. Constant air consumption. If the bleed function were used to replace limit valve C or D in this same circuit, air would exhaust to atmosphere constantly when the machine is idle.
2. A break in the line between the two-way valve and the a port of element 8 converter would cause the output of this element to go on. This has possible safety implications, although not any worse than a three-way valve sticking on or being accidently actuated.
3. Response times may be extended. Depending on the orifice size in element 8, tubing sizes, and the two-way valve's flow, the response time lag may also be uneven. For example, it may take 50 milliseconds for the two- to three-way converter to open once

the two-way valve has been actuated, but it may take 200 milliseconds for the two- or three-way converter to close after the two-way valve has closed.
4. This is one that should be carefully examined—the two- or three-way converter will pulse each time the air supply to the circuit goes on. The output of the converter element will be on until the supply has a chance to build pressure through the orifice in the converter by filling the line to the two-way valves. So, if we converted the three-way start push button in this circuit to a two-way bleed function, the machine would start each time the supply air was applied. This has definite safety implications and should be carefully studied each time a bleed signal is applied.

PRESSURE SENSING (INCREASING)

Replacing limit valve A in Figure 2 is a pressure-sensing valve. This pressure-sensing valve will be actuated when the pressure in the blind end of the cylinder arrives at a given pressure (usually adjustable on the pressure-sensing valve itself). This could be quite valuable in some applications if, for example, the sizes of the parts varied and the clamp cylinder stopped at various points. Pressure sensing also removes the need for locating and mounting the limit valve, so the pressure-sensing valve may also be easier to install.

Again, however, some negative factors should be recognized. They are:

1. Adjustable pressure-sensing valves are often more costly than a limit valve. This, however, can sometimes be offset by their comparable costs of installation.
2. Adjustable pressure-sensing valves can be adjusted too low (punch cylinder advances too soon) or too high (punch cylinder will not advance). Both of these conditions should be studied carefully, especially as they apply to safety, but also as to how they will alter your troubleshooting methods.
3. Still other adjustments can affect the shift point of the pressure-sensing valve. The adjustment of the flow control and the regulator shown in Figure 3 both can affect the pressure-sensing valve. The regulator adjusted too high can cause PS-A to shift too soon, too low—not shift at all. The flow control, if overadjusted to slow the cylinder on the forward stroke, can also cause PS-A to shift too soon. These factors again should be studied carefully for safety problems. In addition, the instructions for the machine must be examined to be sure that the cause-effect

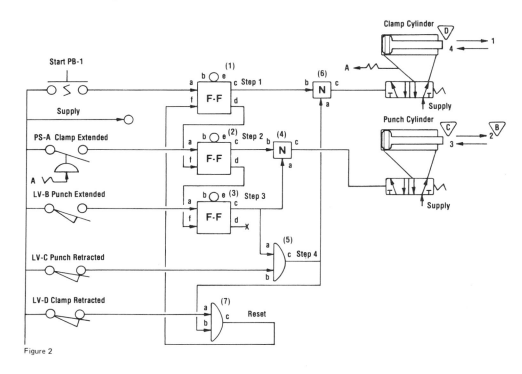

Figure 2

relationship between these seemingly unrelated adjustments is clearly explained. Visual assistance can often be a tremendous help. Notice that Figure 3 shows a gauge on the regulator and on the pressure return line A to help repair personnel coordinate these adjustments.

4. Finally, consider one additional fact carefully. The pressure signal indicated only that the cylinder pressure has reached a given level, not that any movement has taken place. Any obstruction that will cause the cylinder to stop or create a load great enough to build sufficient pressure on the return line (A) will cause this pressure-sensing valve to shift.

PRESSURE SENSING (DECREASING)

Some of the problems associated with pressure sensing are reduced when decreasing pressure sensing can be used. Figure 4 shows decreasing pressure sensing, again replacing limit valve A. The pressure-sensing valve

Pressure Sensing (Decreasing) 129

(PS-A) is now a normally passing valve held nonpassing by the return pressure from A. Notice also that signal a pressure comes from the rod end of the clamp cylinder. This may be hard to visualize at first, but to begin with, full line pressure will be present at A (the cylinder held retracted). When the valve shifts and the cylinder starts to extend, some of this pressure will remain. That is what controls the speed of the cylinder. As the air from the rod end of the cylinder exhausts through the flow control valve, the piston of the cylinder will advance, reducing the volume in the rod end section of the cylinder, in effect, recompressing the air. The pressure from signal line A would be similar to the chart in Figure 5.

The advantages to this type of pressure sensing are that adjustments to flow controls and to the operating pressure of the cylinder have less effect on the adjustment of the pressure-sensing valve. In some cases, a pressure-sensing valve with a fixed shift point of, say, 5 psi decreasing and 20 psi increasing, could be used for these applications, eliminating the potential for misadjustment.

In some circumstances, snap-acting circuit elements can act as pressure-sensing valves. Figure 6 shows Not 8 and And 7 acting as pressure-sensing valves, eliminating limit valves A and D.

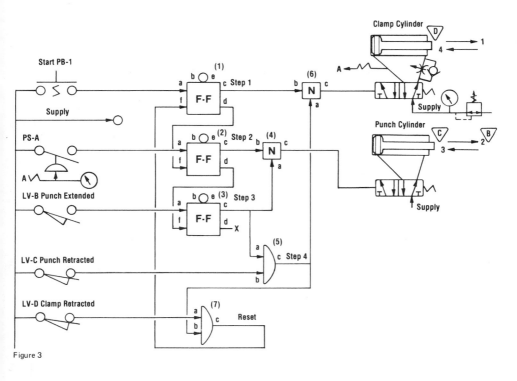

Figure 3

130   Input Variations and Other Special Circumstances

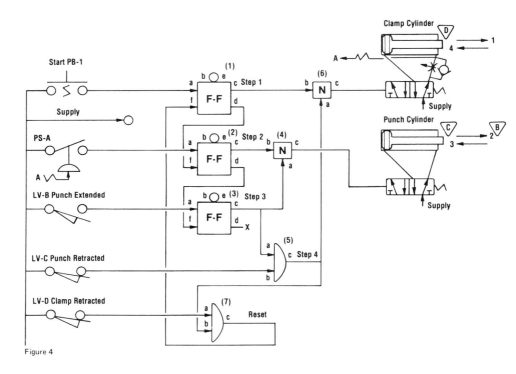

Figure 4

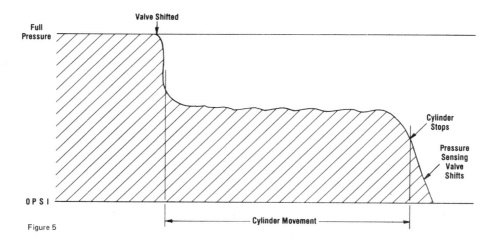

Figure 5

Other Input Modifications 131

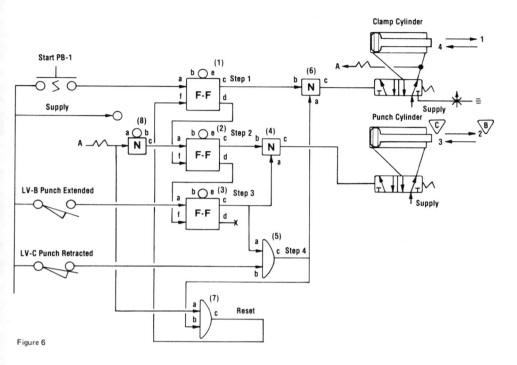

Figure 6

Special circuit elements, some with adjustable shift points called sequencing valves, are also used in place of pressure-sensing valves. With the exception of the one advantage mentioned, the same caution should be applied when using decreasing as with increasing pressure sensing.

OTHER INPUT MODIFICATIONS

Noncontact sensing, electrical interface, vacuum, and other types of input conversions are handled similarly to the conversions already shown. These will be included in the illustrations that follow. The important thing to remember is this: none of the changes up to this point will cause a change in the circuit design method described. Once the signal from these devices is converted to an air signal compatible with our system, it is then judged to be either maintained to the end of the cycle or not maintained to the end of the cycle. Following this analysis, the procedure of design for the sequential circuit remains unchanged.

132   Input Variations and Other Special Circumstances

The Missing Input

This one *does* require a circuit change. "Automated" tools, index tables, and other devices often have an internal circuit. Theoretically, this is supposed to make automation simpler when using these products. In some applications it does—in others it does not. Figure 7 shows drawings of two of these devices. In the automatic drill you have a drill and a motor. A signal at the start will cause the cylinder to advance. As the cylinder advances it releases a valve (internal) which supplies air to the motor (A). When the drill reaches depth, it actuates a limit valve (B) causing the feed cylinder to retract. The cylinder retracts to actuate limit A, turning the motor off. This completes the cycle. Here is the problem. Limit valves A and B are internal; the only signal available is the drill-retracted signal coming from limit valve C.

The automatic index table is a similar problem. Here we have two cylinders which go through a four-step sequence for each index. Again, all of the valving is internal and the only signal that we can use is the C limit valve output, in this case indicating that the shot pin is in position.

There are a couple of ways to approach this problem. The first might be to deautomate those devices. Buy a unit without these internal controls since in this case they seem to be working against you. This is quite often not possible due to the way they are designed to be manufactured. The second is to design a circuit which will accommodate their automated functions. To integrate these devices into a sequential circuit requires that you do two things:

1. You must signal the start input long enough to get them started. Manufacturers' literature will often call this a pulse signal. Actually, this is misleading. What they really are indicating is that the start signal must be removed in order for the unit to complete its automatic function. The automatic drill is a typical example of this. If the start signal is still on when the drill actuates, the B limit valve—the valve on the feed cylinder—cannot shift.
2. You must convert the signal from limit valve C to one which indicates that a cycle has been completed. Remember that this signal is on whenever the drill is retracted.

Figure 8 charts the start signal and the signal from limit valve C. Below this is the circuit that is used to control the drill. Following the diagram and circuit on Figure 8, here is a description of how it functions.

To begin with, flip flop 3 is in the reset condition, being reset at the end of the previous cycle. Limit valve C is signaling the a input of Not 2 and And 4. Not 2 and And 4 outputs are off.

Other Input Modifications   133

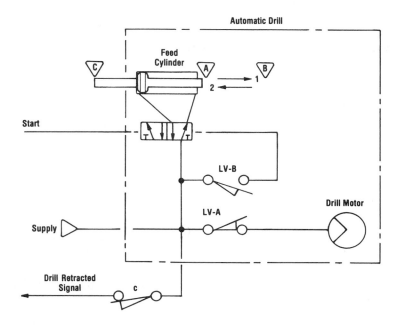

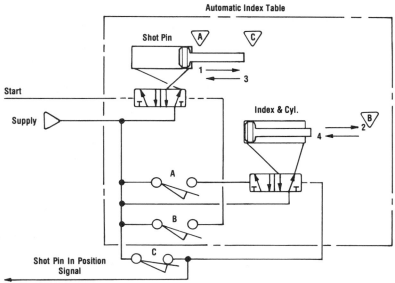

Figure 7

134   Input Variations and Other Special Circumstances

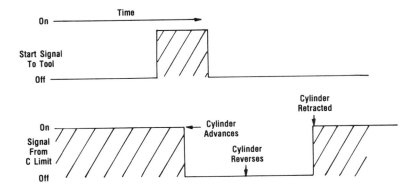

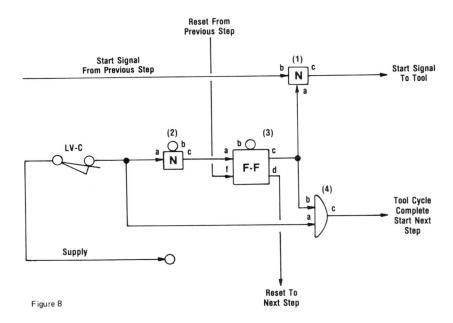

Figure 8

1. A start signal, coming from a previous step in our circuit, goes to the b port of Not 1 and out the c port to start the tool.
2. When the tool starts it advances, releasing limit C. This removes the signal to the a port of Not 2 and And 4. The signal from the c of Not 2 sets flip flop 3.
3. The set output (c) of flip flop 3 signals the a port of Not 1. This cuts off the start signal to the tool. Now note: the start signal to

the tool was *not* a pulse signal. The start signal was *maintained* until a feedback signal from the tool (release of limit C) indicated that it had done its job.

4. With the start signal now off, the tool will complete its cycle and return to again actuate limit valve C. Again, the a ports of Not 2 and And 4 are pressurized. The output of Not 2 goes off; however, flip flop 3 remains set. Now both inputs to And 4 are on. The output from And 4 indicates that the tool cycle is complete.

Figure 9 shows this circuit in application. Here we have a clamp cylinder, a drill, a tap, and a slide cylinder. The part is clamped, a hole is drilled, the slide moves the part to the tapper, and the hole is tapped. The slide then moves the part back to the operator and the part is unclamped. The drill and the tapper are automatic units and the circuit just described is duplicated for each unit (3, 4, 5, 6 and 9, 10, 11, 12).

A variation of the circuit described is required when the input signal is *on* during the automatic cycling period. Figure 10 shows an example of this type of signal produced by an automatic drill. The circuit for this type of signal is shown in Figure 11.

As before, the start signal is generated from a previous step. When the tool-running signal comes on, it sets flip flop 2 and actuates the a port of Not 3. Flip flop 2 cancels the start signal to the tool at Not 1 and signals the b port of Not 3. When the tool-running signal is removed, the a port of Not 3 is released and the tool cycle signal goes on.

There are some cautionary notes that apply to using these circuits. First, be sure you know what "kind" of signal will be produced by these tools. What pressure will it be, and what "pattern" will it produce? Notice that in Figure 7, limit C is supplied by the tool supply. This means that the signal from limit C could be greatly affected by the air consumption of the drill motor, the porting within the tool, the regulated pressure to the tool, and other factors. Tests may be required to be sure that the drill-retracted signal will be compatible with the air circuit. If not, it may be necessary to supply this limit from the control circuit supply rather than the tool supply.

In Figure 10, this problem gets worse. Here, if we find that this signal is not compatible, we have no alternative but to add a device to convert these signals into a usable form. Figure 12 shows the drill and tap circuit with "tool-running" signals. Here also, we have added an additional pressure-sensing valve to correct the signal pressure problem just described. These need not be elaborate adjustable pressure-sensing valves, but can be any pilot valve which can be shifted by the pressure from the tools. In some cases, it will be increasing the pressure to the circuit (amplification) and in

136 Input Variations and Other Special Circumstances

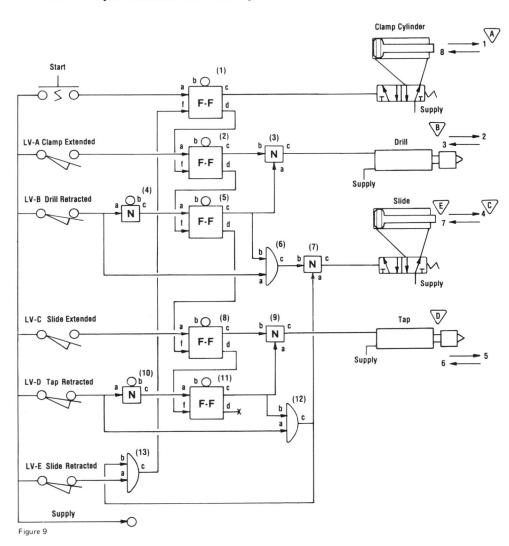

Figure 9

others it will be reducing the signal pressure to the circuit. The valve also serves to isolate and prevent the lubricated air from the tool from getting into the control circuit.

The second cautionary note is this: There is a potential in this circuit for a momentary "tool cycle complete" signal as the tool begins to run. This depends almost entirely on the pressures that the elements were designed to shift. This problem is illustrated in Figure 13. These are caused

Other Input Modifications 137

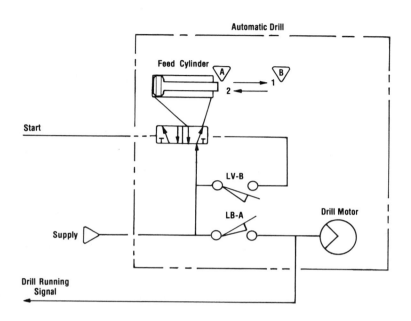

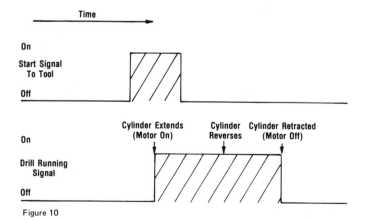

Figure 10

138   Input Variations and Other Special Circumstances

by an "early" shift of the flip flop, either before the signal has been fully exhausted at the And element (as in the case of Figure 8) or before the signal has increased to the pressure needed to close a Not element (Figure 11).

So, whenever you use this circuit with a new brand of controls, be sure to investigate the shift points of these elements. This is not an unusual circuit requirement, so most manufacturers have designed these components to prevent this or will be able to furnish an alternate circuit for this function. If for some reason the problem exists and cannot be solved in this way, a delay can be added to these circuits to ensure that the output pulse will not occur. Figure 14 shows the two circuits with delay elements added.

Here are two applications which involve these circuit combinations. Figure 15 is a combination of an index table and an automatic drill. A part is placed on the index table, clamped (manually), and the start push button is actuated. A series of four holes in the same part are then drilled by indexing and drilling continuously until the part has turned 360 degrees (four indexes followed by four drilling operations). This circuit uses the Not-flip flop-And combination three times: once to indicate the movement of the index table one position (4, 5, and 6); the second to indicate a drill cycle (8, 9, and 10); and a third to indicate a full revolution of the table (13, 14, and 15). Elements 13, 14, and 15 replace the counter shown in our previous example of a continuous operation circuit. These elements are operated by a limit valve, which is actuated only in the start/stop position of the index table, and operate the redirect circuit (11 and 12) to shut the circuit down after the final drill operation has been performed. Notice that flip flop 14 is reset by flip flop 1.

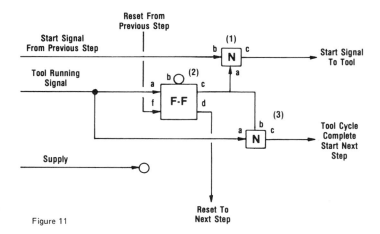

Figure 11

Other Input Modifications 139

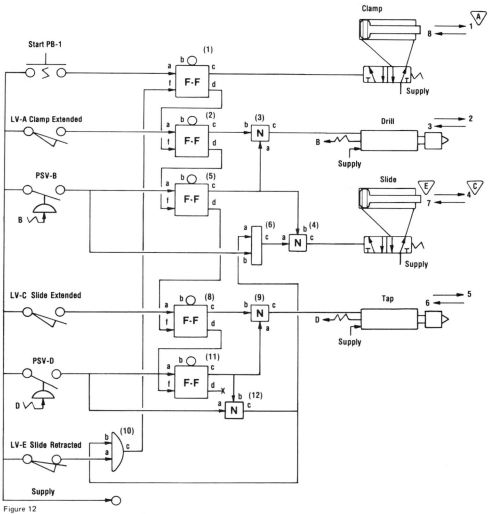

Figure 12

140 Input Variations and Other Special Circumstances

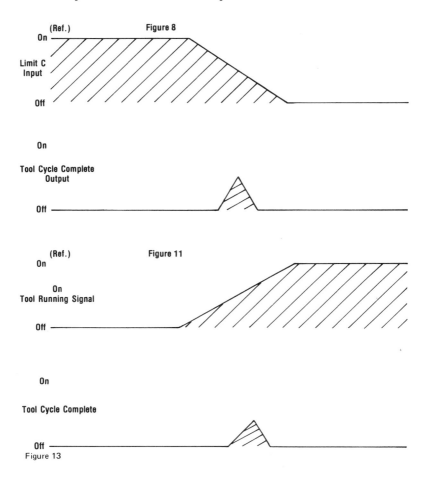

Figure 13

Figure 16 shows still another application of this type of circuit. Here we have a slide, a slide clamp, and an automatic drill. The operator loads and clamps the part manually and hits the start push button. The slide then moves to limit B and stops, the slide clamp extends, and the drill cycles. When the drill has retracted, the slide clamp retracts and the slide again extends, until it again actuates limit valve B. This cycle is repeated until the slide actuates both B and F limits. This time, when the slide clamp retracts, the slide will return to limit valve A and the system will reset.

Notice that this circuit uses the Not-flip flop-And circuit to detect a movement of the slide cylinder from one position to the next. Limit valve

Other Input Modifications   141

B is released and reactuated as the cylinder moves one position. Elements 5, 6, and 7 perform this operation. Elements 11, 12, and 13 monitor the drill cycle. Note also that the b ports of Not 5 and Not 11 are connected to the output of the previous step. With this connection, the circuit can be reset at any time as is required by the abort emergency stop used in this application.

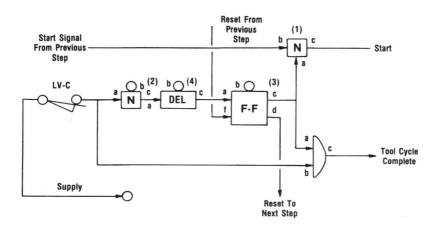

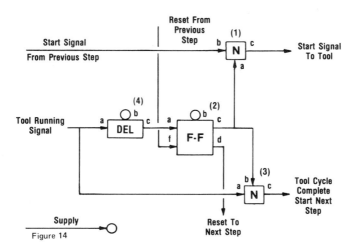
Figure 14

142 Input Variations and Other Special Circumstances

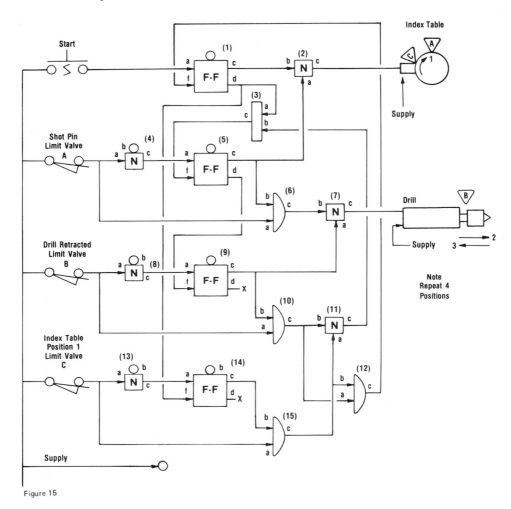

Figure 15

TIMING

Up to now we have covered basic circuits using limit valves, variations of these basic circuits, variations of input devices, and how to compensate for inadequate input signals. Now we will cover the final phase of circuit design—operating control circuits with no input signals based strictly on time.

There are several good reasons for using time as a base for anything

Timing 143

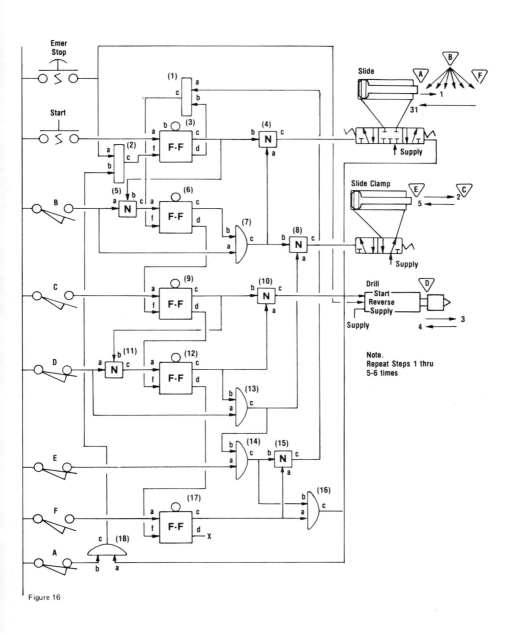

Figure 16

144   Input Variations and Other Special Circumstances

from a single step to an entire circuit program. The step itself could be time based. Holding two parts together long enough for an adhesive to cure would be an example of a pure time base requirement. In other cases, timing may be substituted for a hard-to-measure variable. A cool-down period on a mold could be an example of this. The mold could be opened based on temperature; however, timing is substituted because temperature sensing is more complicated and costly and it was decided that timing would be adequate and safe for this application.

Cost is still a third factor. In many instances, timing is substituted for limit valves, where they could have been used, simply because the application was noncritical and using timing would deliver a machine lower in cost overall.

Applying timing functions to a sequential circuit is a simple process. In Figure 17, we have replaced the limit valves D and B plus the stage elements they would have actuated with time delays 3 and 7. Step 2, the extension of the punch cylinder, is now a time-based function. When the signal is sent to extend the punch cylinder (step 2 on), a time delay (3) is started. When this time delay output goes on, step 3 is created and the punch cylinder retracts. The reset of the circuit is now also time based.

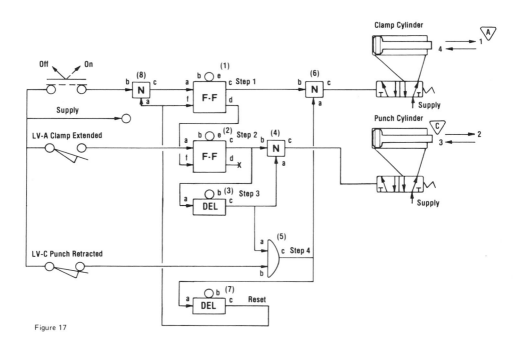

Figure 17

Step 4 starts delay 7. When delay 7 output goes on, the circuit is reset and the system repeats.

Notice that the step outputs are still maintained signals. The outputs of the delays, flip flops, and And elements will be maintained until the system is reset at the end of the cycle.

Figure 18 is a three-cylinder, six-step sequence. With the exception of the start signal, all of the functions of this circuit are time based. Notice that the delay functions are connected in series in the order they are to take place. The output of delay 3 starts delay 5, the output of delay 5 starts delay 6, and so forth. Delay functions should always be connected in series for a sequential circuit. This reduces the adjusted setting for each delay, ensures that the delay adjustment cannot change the sequence, and ensures that all delays are reset before a new cycle can occur. In addition, by resetting the delay function in the same order they occur, we can be sure that the b port of the Not elements are off before the a input is released.

Deciding when a delay function can and cannot be used is very important. Some rules which may help in this decision are listed here:

1. Safety: if an accident can occur because of a misadjustment to or malfunction of a time delay, a delay function cannot, of course, be used. Nonadjustable (fixed) timers are sometimes used in these instances (such as a two-hand anti-tie-down). More often, in the case of sequential circuits, a more positive means of detecting the completion of a step are substituted.

2. To a lesser degree, the dollar consequences of a misadjustment or malfunction are also important. Possible expensive damage to a part or to the machine itself would be reason to consider alternatives.

Maintenance, adjustment, and positioning of these delays are also critical to the dependability and repairability of the control circuit. Be sure you clearly describe the adjustments, what they do, and how they are made, to those who will be making them. The type of adjustment and positioning are also important. Just as in manual controls, you can give adjustments to particular individuals and not to others. Listed below are various types of timers and locations. With each one are general types of applications where they might be used.

Product 1: Dial timers with adjustments that read in seconds, minutes, or hours. These are sometimes called process timers and often include an indicator to show that the timing function has begun and what portion of the delay remains. Process timers are normally used where frequent adjustments are required by the op-

146 Input Variations and Other Special Circumstances

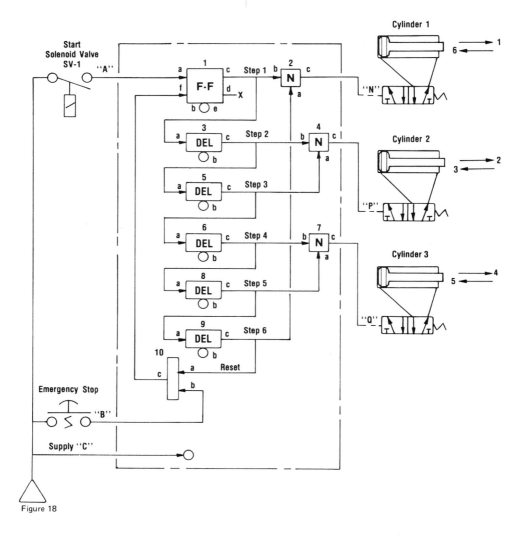

Figure 18

erator and they are located on the operator control panel. The application involving the adhesive cure time, previously mentioned, might be an example of this.

Product 2: Dial timers which have reference numbers (but do not show exact times) are sometimes used for set-up personnel for infrequent but necessary adjustments during changeover from one job to the next. These would normally be mounted in the control enclosure or elsewhere, but away from the operator control station.

Product 3: Screw adjustments and other types of semirestrictive adjustments. Semirestrictive adjustments are used for delays which will seldom need readjustment once in service. Any adjustment would be done by maintenance personnel. These would generally be located in the control enclosure. The delays used in Figures 17 and 18 would probably fall into this category. Both this type and type 2 are sometimes locked in an enclosure with the key given to the proper personnel.

Product 4: Nonadjustable delays are used in safety circuits and other critical functions which must employ time delays.

So, as you can see, the circuit designer does have some control over adjustment and convenience by the type of delay specified and its location.

Additional examples of circuits involving timing, pressure sensing, and other types of input variations are shown in the next chapter.

11
ADDITIONAL CIRCUITS AND APPROACHES

INTRODUCTION

This final chapter is a potpourri of examples. These examples illustrate specific useful circuits in some cases; however, in most they are used to illustrate a useful approach to certain types of applications which you may encounter.

START/STOP CIRCUITS

Figure 1 shows a circuit which could replace the on/off selector valve in a continuous operation circuit. The characteristics of this circuit are as follows.

1. Single-cycle push button produces an 80 millisecond pulse when actuated; auto. mode only.
2. Auto. cycle start sets memory, provides constant start signal; auto. mode only.
3. Auto. cycle stop removes start signal; any mode. Overrides start signal. Cycle completes and stops.
4. Emergency stop removes start signal; any mode. Overrides start signal. Also signals emergency stop to sequential circuit.
5. Auto.-run-memory will also be reset by: (1) auto.-off-manual selector to off or manual, or (2) loss of air supply.

Some of these characteristics may seem redundant; however, in cases where the cost is the same, why not? One factor is important in this type of

Liquid Level Sensing   149

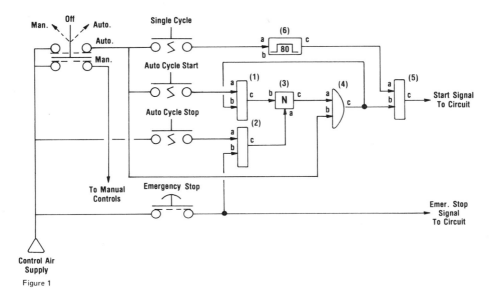

Figure 1

circuit. If you can learn what the operator will expect when making these selections and operating these buttons and then design the circuit to operate in this manner, you will achieve operator acceptance of this machine much more quickly. Sometimes this can be determined by studying the other machines in the department or by asking the operator. In any event, you may be amazed at how much better a machine will function once it has been accepted by the operator.

LIQUID LEVEL SENSING

Air back pressure can sometimes be used for sensing the level of liquids. Figure 2 shows the circuit used for liquid level sensing. The top drawing employs an amplifier with an extremely high gain (approximately 1000 to 1). The lower drawing can function with amplifiers of 100 to 1 or less. The sensing tube will normally touch the liquid, and in some cases will trigger the amplifiers when it is 1 or 2 inches below the surface. This depends on the sensitivity of the amplifiers and the density of the liquid. The sensing tube will, however, normally be kept clean by the flow of air.

Some bubbling normally occurs with direct air back pressure sensing of liquid levels. For applications where this is a problem, floats and other devices are normally used.

150   Additional Circuits and Approaches

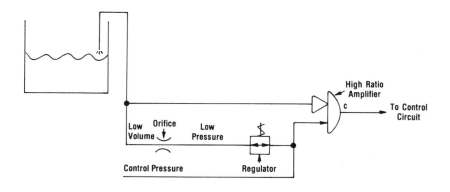

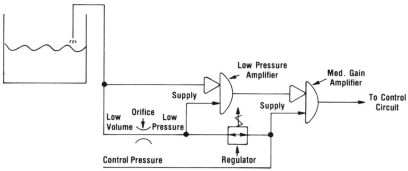

Figure 2

In set-up and testing, an adjustable orifice and regulator are often used. For actual applications, however, these should be replaced with a fixed orifice and regulator wherever possible.

MASTER/SUBCIRCUITS

A common circuit requirement is shown in Figure 3. A master and subcircuit arrangement is required when one portion of a circuit must operate independently of others, for example, an index table with more than one working station. The index table indexes one position. This would be controlled by the master circuit. Then the working stations perform their operations. These working stations operate independently of each other and at the same time. Thus they cannot be included in one long chain of events as with our previous examples. Instead, for a portion of the cycle, the control is given over to several independent control systems, as illustrated in Figure 4.

Design of the master control and the independent control circuits can follow the same rules we have described for sequential control circuits. The trick is to include enough interlocks in the master control to ensure that each one of the subcircuits is actually performing its sequence. Figure 5 shows a master circuit (A) and one of the four subcircuits (B). At first glance, the interlocks between the master and the subcircuits may appear to be sufficient. The master circuit monitors the index and starts the subcircuits. When the subcircuits have completed their sequence, they signal Ands 6, 7, and 8. When all subcircuit cycle complete signals are present, the master circuit resets and signals all of the subcircuits to reset. This

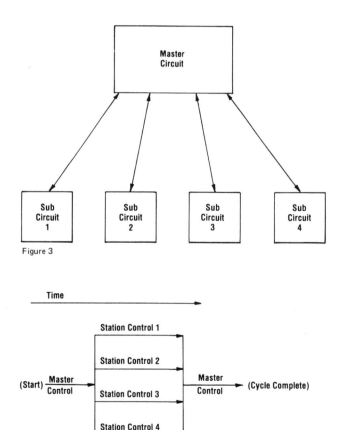

Figure 3

Figure 4

152 Additional Circuits and Approaches

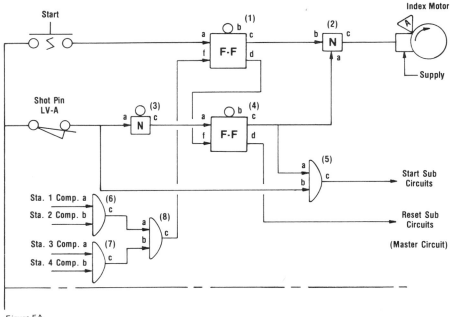

Figure 5A

amount of interlocking may be sufficient for some applications, but in others you will need more.

This circuit does not monitor the reset function of the subcircuit. If, for example, the clamp-extended limit were to remain on in the subcircuit, the station 1 circuit would not reset. The station 1 completion signal would remain on and the machine, with the exception of station 1, would continue to operate.

Figure 6 shows this same circuit with a reset interlock. With this feature, the limit valve function will cause the entire machine to stop just as it would the individual circuit, getting maximum use of the interlocks provided as a result of the sequential method of design used.

Once the design of the master and subcircuits is complete, they may be combined into one complete circuit in the form of hardware, or you may also consider building separate circuits for each function. Consider this carefully, because in some cases much can be gained by building separate circuits. Troubleshooting can be simplified if a problem can be associated with a small group of components. Conversions can be made to individual stations without disturbing the other sections of the control. Occasionally,

Test Circuits   153

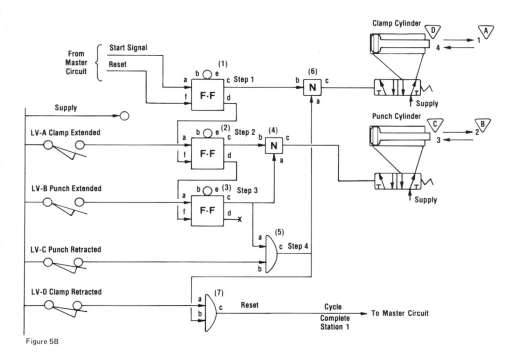

Figure 5B

manual controls can be simplified by building individual circuits. All of these factors being considered, you may decide that separate circuits are an advantage well worth the additional cost.

TEST CIRCUITS

Some circuits start out simple and then get more and more complex. Test circuits seem to fall into this category. Here is an example. We start out with a fairly simple requirement. We have a push-button start. Then a clamp cylinder extends to hold and seal the part. When this is done, we apply 50 psi to the part, close the supply, and check for pressure decay (against time). If the pressure is still above 45 psi when the time is complete, the part is good. If not, it is bad. Simple, right? Here are some additional requirements that are added:

1. Some parts have gross leaks such that we will not achieve 50 psi. We want to stop the test and allow the operator to separate these parts from ones that failed the pressure decay test.

154 Additional Circuits and Approaches

2. For good parts, all phases should operate automatically. For bad parts, the operator must acknowledge and note the type of fault before the clamp cylinder will return.
3. We must reduce the pressure on the part to less than 10 psi before we open the clamp—manual or automatic.

Figure 7 shows the circuit designed for this application. The start push button and the clamp limit valve circuit are pretty standard. Phase 3 is where we begin to do two things at once. When pressure is applied to the part, one of two things can happen first. A good part will cause PSV-1 to close first. A bad part will allow delay 6 to complete. This would cause the fault 1 indicator to go on and abort the cycle. Step 4 also has two options. When delay 8 is complete, the system automatically resets; however, if PSV-2 comes on before delay 8, the fault 2 indicator goes on and this again aborts the cycle. These parallel actions require careful analysis since in many cases they will continue unless stopped by the first result.

Notice that once pressure is achieved, delay 6 is stopped, and also,

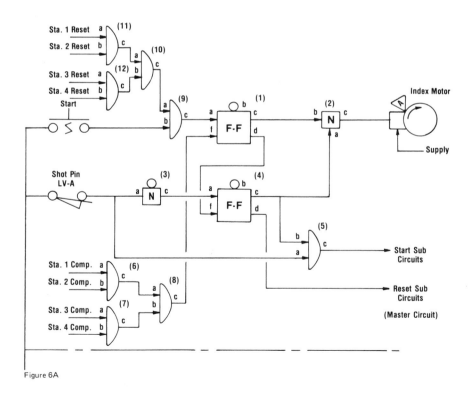

Figure 6A

Conclusion 155

once delay 8 is complete the signal from PSV-2 is prevented from actuating flip flop 12. Circuits of this type require skill in the design of both sequential and combinational circuits. They also offer some of the best opportunities to innovate in the design of a circuit.

CONCLUSION

In conclusion, the variety of circuits which you may encounter is endless, and additional illustrations probably pointless. The attitudes and resulting methods you develop in designing these circuits seem to be the most important factors in the success or failure of these designs. Here is a list of attitudes and methods which we have observed that enhance your chances for success in circuit design.

> Thorough investigation of the circuit requirements initially and an open mind to changes in these requirements as they occur. And they will, even after your design is complete. Don't hesitate to start over from the beginning as "add ons" often become problems. The final circuit with the modifications should be as good as

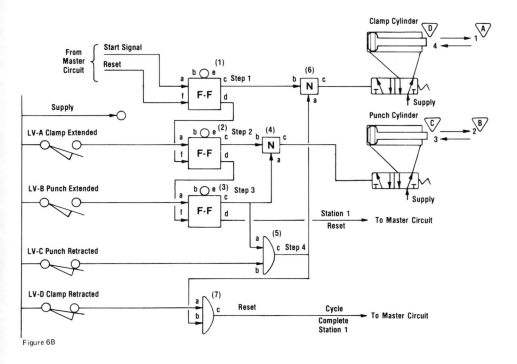

Figure 6B

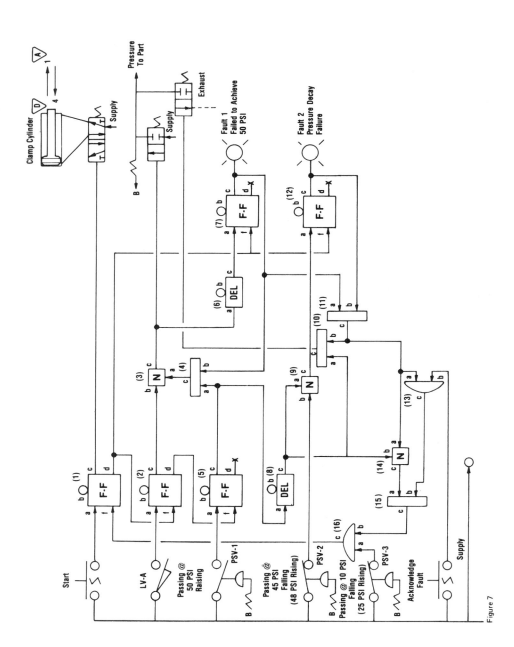

Figure 7

Conclusion   157

one which was designed with all of the facts at hand initially. You first make it work, then you reduce the price. You will find it very difficult to design a system and add up the price at the same time. Once a functioning circuit has been designed, however, the price can often be reduced.

Remember the basic requirements of a good circuit design:

1. Function
2. Safety
3. Reliability
4. Repairability
5. Cost

Consider these carefully in design. Review and test your circuit with these requirements in mind.

APPENDIX A
INPUT SYMBOLS
GENERAL—USE FOR ALL DIAGRAMMING—ATTACHED OR DETACHED METHODS

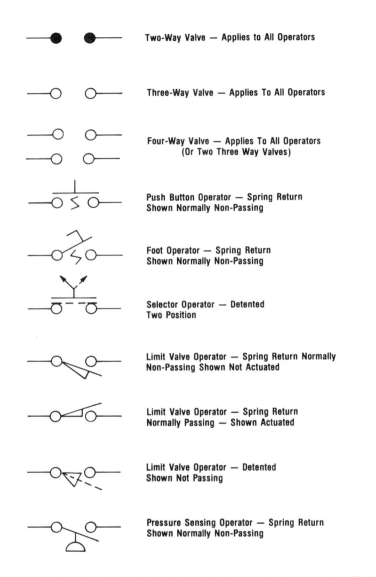

For Additional Symbols and Details For Their Use See ANSI Standard B93.38-1976 (R-1981)

APPENDIX B
CIRCUIT SYMBOLS

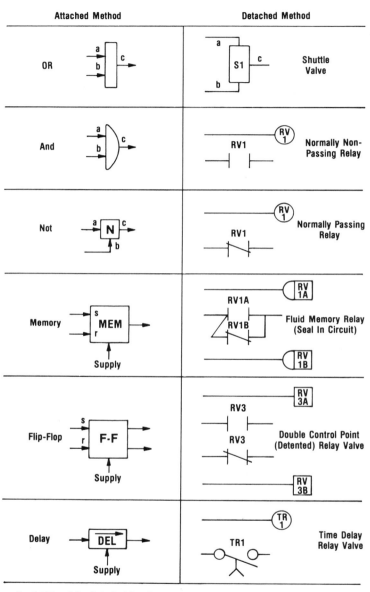

For Additional Symbols And Details For Their Use See ANSI Standard B93.38-1976 R 1981

APPENDIX C
OUTPUT SYMBOLS

POWER VALVE CONTROL POINTS — Use With Detached Method Or Where Power Valves Are Not Connected To Circuit

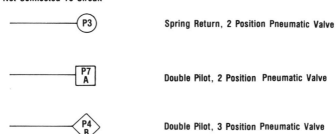

Spring Return, 2 Position Pneumatic Valve

Double Pilot, 2 Position Pneumatic Valve

Double Pilot, 3 Position Pneumatic Valve

For Additional Symbols And Details For Their Use See ANSI Standard B93.38 — 1976 (R-1981)

POWER VALVES, CYLINDERS AND MOTORS

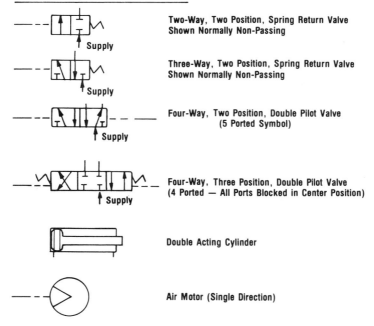

Two-Way, Two Position, Spring Return Valve Shown Normally Non-Passing

Three-Way, Two Position, Spring Return Valve Shown Normally Non-Passing

Four-Way, Two Position, Double Pilot Valve (5 Ported Symbol)

Four-Way, Three Position, Double Pilot Valve (4 Ported — All Ports Blocked in Center Position)

Double Acting Cylinder

Air Motor (Single Direction)

For Additional Symbols and Details For Their Use See American National Standard ANSI/Y32.10

INDEX

Boolean algebra, 33–34

Circuit:
  automatic drill, 132–141
  automatic index table, 132–141
  binary flip flop, 65–66
  combinational logic, 79–80
  continuous operation, 120–121
  contraction, 115–117
  design practices, 80–82
  double sequence, 116–119
  driven sequence, 121–123
  equivalence, 40–41
  expansion, 112–113
  implication, 42–43
  master/sub, 150–155
  mixing, 113–115
  nonequivalence, 40–41
  nonimplication, 41–43
  oscillator, 62–63
  repeater, 123–124
  sequential, 79–80
    design method, 83–94
  shift register, 66–68
  signal standardizer, 60

[Circuit]
  start/stop, 148–149
  test, 153–156
  two-hand anti-tie-down, 60–62
Controls, manual, 104–110

DeMorgan's theorem, 36–37
Devices:
  analog, 9
  binary, 8
  manual input, 11–15
  mechanical input, 17–21
  operation sensing, 9
  operator control, 9
  power, 9
Diagramming methods:
  attached, 4
  detached, 3

Elements (components):
  amplifier, 25, 72
  And, 32–33
  delay, 46
  flip flop, 57–59
  interface, 70–78

[Elements (components)]
   air to electrical, 70–71
   air to hydraulic, 77
   air to vacuum, 77–78
   electrical to air, 70–71
   flow reduction and increase, 73, 75
   high to low pressure, 72
   hydraulic to air, 77–78
   low to high pressure, 72–73
   three-way to two-way, 74–75
   two-way to three-way, 74–75, 125–126
   vacuum to air, 77–78
   Not, 36–42
   Or, 30–32
   pulse, 46–48
   timer, 46

Functions:
   bleed (*see* Element, two-way to three-way)
   inhibit (*see* Element, Not)
   inversion (*see* Element, Not)
   memory, 49–59
   Nand, 40–41
   Nor, 40–41
   pulse inverted, 47–48
   R-S-T flip flop, 66
   sequential timing, 142–146
   snap-action, 2, 44
   timing in, 47
   timing in inverted, 47

[Functions]
   timing in and out, 47–48
   timing in and out inverted, 47–48
   timing out, 47, 63
   timing out inverted, 47

Indicators, 15–16

Sensors:
   air jet (*see* Sensors, noncontact)
   interruptable gap, 26–27
   liquid level, 149–150
   noncontact, 23–28
   pressure build-up, 21–23, 127–128
   pressure decay, 21–23, 128–131
Symbols, composite, 66–69
Systems:
   air control, 9
   emergency stop, 97–107
   moving-part logic, 2

Valves:
   check, 44
   limit, 17–20
   poppet, 29
   power, 1
   push button, 11–13
   selector, 11, 14
   spool, 29
   three-way, 8
   two-way, 20–21